AND

SPHERELAND *A Fantasy About*

Curved Spaces and an Expanding Universe

by DIONYS BURGER

TRANSLATED FROM THE DUTCH BY Cornelie J. Rheinboldt

FOREWORD BY Isaac Asimov

Quill

A HarperResource Book
An Imprint of HarperCollins *Publishers*

FIRST QUILL/HARPERRESOURCE edition published in 2001
FIRST HARPERPERENNIAL EDITION published 1994

The Library of Congress has catalogued the previous edition of *Flatland* as follows:

Abbott, Edwin Abbott, 1838-1926.
　　Flatland : a romance of many dimensions.

　　(Everyday handbook; EH/573]
　　Reprint of: 5th ed., rev., 1963; with forward by Isaac Asimov
　　1. Fourth dimension. I. Title.
　　[QA699.A13 1983] 530.1'1　　　　　　　　　82-48824
　　ISBN 0-06-463573-2(pbk.)

The Library of Congress has catalogued the previous edition of *Sphereland* as follows:

Burger, Dionys.
　　Sphererland: a fantasy about curved spaces and an expanding universe.

　　(Everyday handbook)
　　Translation of: Bolland.
　　1. Fourth dimension. 2. Sphere. 3. Expanding universe. I. Title.
　　QA699.B813 1983 530.1'1　　　　　　　　82-48828
　　ISBN 0-06-463574-0(pbk.)

ISBN 0-06-273276-5
　　04 05 RRD H 20 19

Contents

vi

EINSTEIN HAPPENED

by Isaac Asimov

Why was it necessary to write a sequel to *Flatland*?

Flatland, published originally about 1880, is a charming geometric fantasy which, in the guise of dealing with living, thinking creatures introduces the reader, painlessly, to the mysteries of dimensional thinking. —Even to that all-but-ungraspable concept, the fourth dimension.

But surely geometric fantasies are not subject to obsolescence. Triangles, squares and spheres are today what they were in 1880 and will remain so indefinitely into the future. Why, then, continue the story? What more is there to say?

For one thing, the original *Flatland* was not about mathematics only. It also described a society in which casual assumptions were made concerning class distinctions and, in particular, women. The Victorian convention of women as a quite inferior form of life was accepted without question.

In the twentieth century, this view, both insulting and injurious, to say nothing of being untrue, could not be accepted, and Dionys Burger, the Dutch mathematician who published *Sphereland* in 1960, went to some pains to neutralize it. (Good for him!)

But there was more. Even the mathematics needed additions, for though a square remains a square, subtle changes in our understanding of what squares may be like take place as we gain a better understanding of the Universe.

Let me give you an example. The Earth we live on, in general, looks flat. It is lumpy and bumpy, of course, with hills and ravines, but there doesn't seem to be any general slant. If the unevennesses are averaged out, everything would be flat. At least so our eyes tell us. And since the Earth *looks* flat, the ancients thought, with considerable justification, that it *was* flat.

Today, however, we know that the Earth is spherical. Its surface is curved. The Earth looks just as flat as ever. Nothing has changed as far as our eyes are concerned, but our *understanding* of the Earth as a whole has changed. We know that the surface is curved, but so gently, so unnoticeably, that it still looks flat. Nevertheless, we can't say it *is* flat, because if we travel large distances then a guiding map drawn on the assumption that the Earth is flat will mislead us. Only a globe, or a map that is flat but takes the globularity of the Earth into account will lead us aright.

Well, the Universe isn't flat, either. If it were flat, then a ray of light traveling through a vacuum would move in a perfectly straight path. It would be as close to a perfectly straight line as we can imagine. And if we caused such a ray of light to produce a visible path (as by making it travel through foggy air that would scatter its light slightly all along its progress), that path would certainly *look* straight, just as the Earth certainly *looks* flat.

Just the same the path of the ray of light is not *perfectly* flat. It deviates slightly, *very* slightly, from the straight line. It deviates far less than the surface of the Earth deviates from flatness.

Of course, the deviation builds up. If we were to imagine the path of the ray of light made very long indeed—from here to some distant star, for instance—then it would be quite clear that it did not follow a perfectly straight line. A map of the Universe based on light traveling in a perfectly straight line would not be accurate and if we tried to use it as a guide to travel great distances, we would find ourselves led totally astray—just as we would go astray if we tried to use a map of a flat Earth in order to travel from the United States to New Zealand.

Who told us that the Universe was not flat, but curved?

It was a German-born scientist, Albert Einstein, Between 1905 and 1916, he worked out a totally new way of looking at the Universe, a way that, at first glance, seemed very complicated and "against common sense." Part of it was that light did not travel in straight lines, but followed paths that had very slight curves to them.

Just the same, Einstein's view, however nonsensical it may have seemed to people unprepared for it, made interesting predictions that turned out to be true—that it was impossible to go faster than light in a vacuum, that certain strange changes took place as you approached the speed of light, that mass and energy could be interchanged, and so on.

Of course, it was difficult to accept this, but that's not surprising. Imagine how difficult it must have been for people to believe the Earth was a sphere when they could "see with their own eyes" that it was flat.

When we think of a square ordinarily, we think of it as bounded by four "straight lines." But those straight lines do not really exist in our Universe. When we draw a straight line, it is actually very slightly curved, because it follows the curve of the Universe just as a ray of light does. The curve is so slight that, under all ordinary conditions we can ignore it, but there are scientific situations today in which we have to take such curvatures into account, or we will never understand why the Universe behaves as it does.

That is why it was necessary to write *Sphereland* as a sequel to *Flatland*. In between *Flatland* in 1880, and *Sphereland* in 1960, Einstein happened, and suddenly we were forced to realize that all those straight lines we have always dealt with so confidently were a little more complicated than we thought.

Sphereland, in its way, then, is a geometric introduction to the Einsteinian Universe.

Fear not, however. It contains no difficult mathematics and it won't sprain your understanding. It remains just what *Flatland* was to begin with—a pleasant fantasy. You will have no sensation of "learning" whatsoever, but you will learn just the same. You won't be able to help it. And when you are done with *Sphereland* you will find that someday, if you should happen to encounter the Einsteinian view, it will be that much less difficult to grasp, thanks to what you have picked up in this book.

A Look at FLATLAND *A Fantasy*
About the Fourth Dimension

by A SQUARE

1 Flatland and Its Inhabitants

Visualize a flat extended plane in which two-dimensional geometric figures—somewhat like shadows, but hard and with bright, shiny borders—can move in all directions, this is Flatland and its inhabitants.

The country has a weak gravitational force in a direction called south. The opposite direction is considered to be north and in between, on either side, are east and west. That force enables the inhabitants to orientate themselves.

In certain moderate zones, however, the force is small. While the "delicate" female can sense it very easily, the cruder male occasionally has trouble with it. For example, someone traveling well outside of the inhabited areas may lose his sense of direction. If this should happen, he will have to wait for a rainstorm, since the rain always comes from the north.

The houses are built with this in mind. They are pentagonal. R OO F is the roof against the rain. On the left is

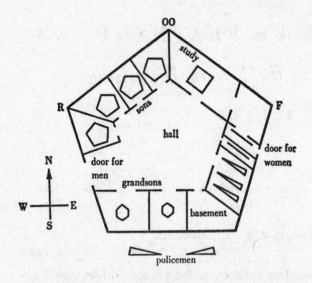

The house of the Square. The bedroom of the man of the house is next to the study. The woman of the house has her own entrance, the Door for Women. Next to it is the daughter's bedroom, and three servants are in the kitchen.

a wide door for men, on the right a narrow one for the women—who are much slimmer, as we shall see.

No windows are needed in private homes because there is light everywhere in Flatland. No one knows where it originates. Perhaps outside the plane, in the space of three dimensions? Square and triangular houses are not permitted because their sharp corners would be dangerous to passersby. We must not forget that vision in Flatland is rather poor since its inhabitants never see more than a line and are barely able to distinguish an angle. The sides of inanimate objects give out very little light and are even more difficult to see. Only military installations such as forts, powder magazines, barracks, and certain government buildings are allowed to have sharper corners, if they are located on grounds not open to the public.

The adult inhabitants are about 11 inches long. The women have the shape of a straight line of almost no width; the men are triangles or polygons. The least developed men are formed like very sharply pointed isosceles triangles with a base of not more than an eighth of an inch and sides of approximately 11 inches each. Consequently, their top angle is very small, and since it contains the brains (and is therefore also known as the "brain angle"), it's clear that we have to do here with the intellectually less gifted.

As each generation succeeds the last, the top angle increases, providing more space for brains. With good behavior (and not otherwise) each successive generation comes to have a top angle half a degree larger than the previous one. Misbehavior or worse can cause an individual to slide down again one or more steps on the social scale.

Whenever a vertical angle of 60° has been achieved, it is first verified by a Sanitary and Social Board established especially for the purpose. When the offspring is certified as an equilateral, it is taken away from the parents and brought up by a childless couple of the Equilateral group.

Development now proceeds very rapidly. The sons of an Equilateral are Squares, their sons in turn are Pentagons (equilateral), then Hexagons, etc., until the number of sides has become so great that the creature resembles a circle. It is then allowed to call itself Circle and admitted to the

Priest class to which all dignitaries belong. This is headed by the Supreme Circle.

The Equilaterals make up the middle class, composed of shopkeepers, merchants, and clerks. The "gentlemen," officials and scholars, are Squares, Pentagons, and Hexagons. They already rank among the nobility, the Polygons (with many angles), who regard themselves as being well above the common workers' class of the Isosceles.

The extremely underdeveloped ones are constantly at war with each other and fights between these sharp-angled creatures are often bloody. The Circles look on this with considerable satisfaction because it lessens the dangers of overpopulation and dangerous rebellions. Nevertheless, it has happened more than once that the lower classes have revolted against the administration. Since the most dangerous individuals with the sharpest vertical angles are also the ones with the smallest brains, such a rebellion is always led by the less sharply angled Isosceles. This fact was turned to good use when the chief rebel leaders were made to undergo an operation in the hospital increasing their vertical angle — which was already close to 60° — to that very value, so that they could be admitted to the higher ranks. Other leaders were also lured to the hospital where they very treacherously were either made prisoners for life or, if they did not go along with that peacefully, were simply killed. The remaining rebels, deprived of every form of leadership, were egged on to fight each other so that the rebellion ended in a mutual murder party.

The females are line-shaped. It is impossible to determine

from their appearance to what caste they belong and their lineage is therefore controlled with care. An upper-class male will not want to marry a woman from a lower rank for fear that his backsliding will show itself in his progeny. Curiously enough, Polygons seem to take the matter more casually. Convinced as they are of their own excellence, they dare to marry the girl of their choice without paying much attention to pedigrees and certificates.

Generally speaking, the women are unhappy creatures. Their very tiny top lets them have very few brains. Because they are needle-sharp, a collision with them is instantly fatal. Laws have therefore been enacted which they are obliged to follow. Each house has a wide door for men and a little narrow one for women, who are allowed to use only the latter. On the street, women are required to keep up a constant warning cry. Failure to do this means the death penalty. It is further decreed that every woman suffering from St. Vitus's dance, fits, or a chronic cold accompanied by violent sneezing, or any other disease involving involuntary motions, must be destroyed immediately.

In certain states women are forbidden to go out on the streets at all. They must remain indoors all their lives except on certain holidays. But this rule has proved unsatisfactory because continuous confinement makes the victims extremely irritable, thereby causing many more marital fights than usual, of course with fatal consequences.

All this curtailing legislation was enacted on behalf of the women as well as the men. Walking backward, for example, a woman can accidentally spear someone and, unable to withdraw her barbed posterior from the squirming victim, can also break into pieces herself. A woman's endpoint is particularly dangerous. First of all, she cannot see what is happening on that side, and furthermore her endpoint is dark, increasing the danger of unexpected collisions. The frontal point, which has an organ functioning as both mouth and eye, can always be distinguished easily as the point that sparkles. Many countries have therefore decreed that a woman must keep her endpoint in constant motion. This commendable custom has been given an extra boost by fashion. The ladies from the higher ranks, in particular, are able to make a rhythmic wagging movement with exceptional grace and all the other women try to copy them as much as possible—still, it is always easy to recognize a lady of the aristocracy by her elegance in this respect.

At home the woman presents a constant danger. To make her angry is synonymous with suicide. The women's chamber, entered through the already mentioned women's door, is so narrow that she cannot turn around in it. The husband can enter via a side door, which he can also use to hurry out again if necessary. This type of construction forestalls a great many problems of possibly fatal consequences. Among the lower classes violent quarrels occur frequently, but the dangers are much the same for both sexes because the upper angle of the men is a fair match in lethal power for the sharp extremities of the women.

In the refined societies of Polygons and Circles, it is the custom for the woman to keep her eye and mouth constantly directed toward her lord and master. This adds to the general safety, because one does not collide as easily with the clearly visible front end of a woman as with her posterior. It does create other problems, however, for having his wife's penetrating eye and especially her constantly chattering mouth aimed at him all the time can drive a man to distraction.

The inhabitants of Flatland never see more than a line, and you may wonder how they manage to recognize anyone! First of all, they use the sense of hearing. The voice of the highly developed Flatlander sounds more aristocratic than that of the common man. But this is not entirely reliable, and besides, some Equilaterals can do a very unfair imitation of a Polygon's voice.

The touch system is better. The introduction of one person to another is done among the ordinary ranks with the customary formula: "May I ask you to feel Mr. X and to be felt by him?" In a shortened version one simply says: "May I feel you, Mr. X?" A rather strange expression, of course, but everyone understands it right away.

"Feeling" means carefully touching one of the angles. As soon as they are of ordinary public-school age, children begin training in this and an adult, after one simple touch, can usually tell the size of an angle pretty accurately.

But recognition can also be accomplished face to face. Since a very thin mist prevails everywhere in Flatland, the more distant objects are always hazy. When one comes face to face with someone else's angle, the sides leading away from it become less clear. This makes it possible, after much practice, to estimate the size of the angle quite accurately.

The feel method is used primarily by the lower ranks of

the population, the view method by the higher ranks. In the exclusive schools of the upper classes, the latter is taught from the very first year.

The life of society in general rests on the regularity of its citizens. Occasionally it happens that a very irregular child appears. If the deviation is not too extreme, some pushing and pulling, done in the clinics, can do much to improve the shape. If the irregularity is too great, the only resort is to kill the misformed creature painlessly. The structure of the entire society would otherwise be disrupted by the existence of such a creature, who might look like a pentagon when seen from one side, while having a much sharper angle on the other. Whether these monsters are really criminals from the moment of birth, or whether they become criminal because of having been ridiculed, rejected, and disdained from childhood on, is an unanswered question: but they constitute a danger for society at large and therefore cannot be tolerated in it.

The state is ruled by the "Circles," who are really polygons with sometimes as many as several hundred sides. It would be difficult to determine the exact size of their angle, even by very careful feeling. The natural law that such a large number of sides denotes distinguished ancestry does not always hold true, because a practice has crept into the highest ranks of the Polygons of having their children, when barely a month old, reshaped in the state clinic. The frame is broken in many places, which increases the many-sidedness. This operation is exceptionally dangerous—only a small percentage of the children survive it. But parental vanity drives the aristocrats to subject nearly all their children to it. The administration tolerates this since the individual's fertility has been lessened with the increase of his sides and if not enough Circles are born, more must be made.

To us three-dimensional creatures, life in a flat plane seems terribly dull. The entire panorama a Flatlander sees is line-shaped. Also, there is very little color difference, as far as we know. This was different once, as the Square tells us. Chromatistes, a Pentagon who invented the art of painting, launched a color fashion which became extremely popular. All the Flatlanders began to paint their sides in different colors. This not only made life more agreeable, it made mutual recognition easier. There was no longer any need to teach the art of angle viewing. The higher ranks had always excelled in this skill and that superiority over the less-developed creatures suffered a painful setback. This in turn led to heightened jealousy and friction among the classes, together with an increasing amount of deceit. For example, a simple Isosceles would sometimes paint himself with many colors, making others see him as a respectable Pentagon. As was predictable, the bubble finally burst. The Circles and Polygons, who had their doubts about the outcome of an open battle with the sharp-angled ones, resorted to an underhanded plot. Under the guise of introducing a very democratic color law, they held a council meeting in which the Head Circle, named Pantocyclus, addressed the crowd. By emphasizing the disadvantages of the new law he overruled the hesitant middle class. The Isosceles who were already close to being Equilaterals were no longer particularly interested in equalization of all ranks and declared themselves to be against democratization. When things had reached this stage, a prearranged signal was given and a terrible blood-bath took place. The ruling parties ordered a charge on the Isosceles by an army of very sharp-angled soldiers, who knew little or nothing about the issue, and by a special regiment of orphan girls. No second charge was needed, because the confusion in the ranks of the attacked was such that they thrust around blindly, inflicting countless casualties on themselves. The fact that the very sharp-angled soldiers fighting for the aristocrats had also decimated their own ranks in the confusion of the battle was not unwelcome to the Circles.

Since that time, painting has been abolished and the ban strictly enforced. Peace has returned and the world of Flatland is ruled strategically as of old by the Circle group.

2   Dream Vision of Lineland

On the next-to-the-last evening of the year 2000 of the Flatland era, the Square, who has been explaining the particularities of Flatland to us, had a vision. He saw a line-shaped land, in other words, a one-dimensional world.

The Square visits Lineland.

The creatures living on this endlessly long line were little lines. They moved back and forth along it and of course were not able to pass each other.

Something like a rush or a roar would go out from the entire line, changing occasionally into a chirping or a warbling, then suddenly stopping again as everything became still once more.

One line was longer than the others, and naturally thinking this was a woman, he addressed her as such and asked what the excitement was all about.

The individual he had addressed said crisply: "I am not a woman. I am the ruler of this world, the king of Lineland."

This potentate could of course not conceive of anything existing outside his own line-shaped world. As far as he knew, his line made up the entire existing space. He could not imagine where this stranger, suddenly looming in front of him, had come from since he, the king, could have no concept of a direction perpendicular to his own world.

The little lines in his world are the men, the dots the women. The visitor saw four men standing on either side of the king, followed by a shorter little line, a boy, in turn followed by four women. They could all move back and forth over a very short piece of their world, and—their two eyes being located with their mouths at each extremity—no one ever saw anything but a dot.

All this seemed terribly dreary to our Square and he wondered whether any sort of family life was possible here. With this in mind he asked the king a somewhat personal question about the state of his family's health.

"Oh," the king replied, "my wives and the children are all hale and hearty."

This surprised the Square very much because only men were standing on either side of the king. He said that he did not understand how His Majesty could ever see his wives or approach them, and that would seem to be quite essential for a marriage and children.

The king found that a little silly. "The birth of children," he explained, "is too important to depend on something as haphazard as contact"; and he went on to tell how each man has two mouths or voices, a bass at one end and a tenor at the other.

"I wouldn't bother telling you this," he said, "if I hadn't noted in the course of our conversation that I did not hear your tenor voice."

The Square explained that he had only one voice, which led the king to observe: "That confirms my impression that you are not a man, but a female monstrosity with a bass voice."

The women in Lineland only have one voice, you see, and it is either a soprano or a contralto. A marriage materializes through the harmonizing of a male's bass and tenor voices with the soprano and contralto of two females, which in turn results in the always simultaneous births of two girls and a boy. By means of this arrangement on the part of nature, the balance of the sexes is always preserved in Lineland.

Next, the Square asked the king how he was able to recognize his subjects, and the latter immediately proceeded to show how he could call his wives, who at that time were each

596 miles 397 yards 33 inches away from him — one to the north, the other to the south.

The king put both his voices into action. On hearing them the women took note of the short time difference between the voices, which indicated to them, after a brief calculation based on the velocity of sound, that the distance between the king's two mouths was 6.083 inches. As the king explained, the calculation was not really necessary for his wives because they had already known their husband for a long time, but in this way one can estimate a fellow Linelander's length quite accurately in the course of conversation.

Then the Square wanted to explain to the king how very limited his life in the line-shaped world was. He pointed out how a man with his type of vision could never distinguish anything more than a dot while he, a Flatlander, could see the difference between a dot and a line. Of course the king was unable to understand this. "How can you claim," he asked, "to see a line, in other words, the inside of a man?"

The Square said that he could see four men, a boy, and then four women on either side of the king, but this did not impress His Majesty in the least, because even the smallest boy in his country would know that.

The Square then took a different tack and explained how, in addition to the north-south one, there was still another direction. Naturally, he could not satisfy the king's request to show it to him. To see it, the king would have to step outside his line.

"Step outside my line?" the king cried out. "Outside my world? Outside of space?"

"Yes," the Square answered, "outside your world, outside of your space, because your space is not the real space. The real one is a plane and yours is only a line."

It goes without saying that this went beyond the king's comprehension altogether. A further suggestion by the Square that the king should try, just once, to move in the direction of his side rather than that of his extremities, did nothing to clarify the matter.

"How can anyone move in the direction of his own inside?" the king asked. Now the Square was forced to resort to action. He pushed himself in front of and past the king all the way through Lineland with the result that the king kept seeing him reappear and disappear. But the king regarded it as little more than a magic trick beyond his comprehension.

When the Square gave him a sideways shove in his inside, the king did experience a painful sensation in his stomach, but this still failed to provide him with insight into the existence of a direction perpendicular to his line.

Then the Square got angry and started calling the king stupid and simple-minded, and when he again pushed his way through the king's world, and the latter, now thoroughly aroused and furious, wanted to ram the nasty intruder, the Square awakened in time from his strange dream.

3   The Visit of the Sphere

The following day was the last day of the year 2000. On the stroke of twelve the year 2001 would dawn and with it the third millennium of the Flatlander's era. The Square, who is known to have been a great mathematician, often taught his grandson geometry. He really enjoyed doing this, for the boy, who had been living with him since the time he lost his parents when still very young, was a purely constructed Hexagon with a clear intellect and an exceptional talent for mathematics. And that is how the Square happened, late in the evening of this last day of the millennium, to be teaching the boy about the connection between arithmetic and geometry. As teaching material a number of squares with sides of 1 inch each had been placed on the ground. He now arranged nine of them into a larger square, each side of which was 3 inches long, and pointed out that it was very easy to calculate the number of square inches of the square, even though the inside of a square can never be seen. "In fact," said the Square, "the inside amounts to 3^2 or 9 square inches."

The small Hexagon thought about this for a little while and then said: "But Grandfather, you have also taught me to raise numbers to the third power. Then 3^3 probably has a geometric meaning too."

The grandfather explained how that could not be. "Whenever a point is moved over a distance 3, that point describes a line of length 3. If now the line is moved in a direction perpendicular to itself over a distance 3, a square results which can be represented by 3^2."

"Fine," the grandson said. "If the square, which is 3^2, is moved over a distance 3 in a direction which I do not see, something has to come into being which can be represented by 3^3."

"But that is nothing, after all!" the Square said.

"I can't see it either," the boy answered, "but I am merely continuing my calculations in the same way."

"In geometry we must always stick to what is logically possible," the grandfather stated and sent the boy to bed.

"The boy talks nonsense," the Square then said, more to himself than to his wife, "a stupid boy."

"He is not stupid," his wife bounced back. "The boy has a bright mind. And you'd do well to remember that, according to the law, you as a Square owe respect to a Hexagon, even if he is your own grandson."

"What he said is still nonsense," the Square muttered again.

"No, it isn't! The boy is certainly not stupid. What he said wasn't nonsense at all!"

Who said that? Where did that strange voice come from? There was no one to be seen, but suddenly a dot materialized which changed into a small circle. That little circle grew bigger and bigger until it had a center line of about 12 inches.

For a moment the Square and his wife were speechless. It is strange enough to be visited at home by a Circle, but when such a dignitary makes his entrance in such an unaccountable way — that is extremely odd.

"I should like to talk with your husband alone. I have a message for him," the visitor stated.

The wife, who, like all wives in Flatland, was very humble and obedient, took her leave and went to her room. It was just twelve o'clock; the new millennium had dawned.

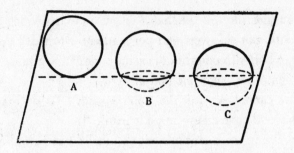

The Sphere visits Flatland.

"I am not a Circle," the stranger explained, "or rather, I am a more complete Circle than you can visualize in Flatland. I am many Circles in one."

"Where do you come from?" the Square wanted to know.

"From space," was the answer.

"But I just saw you enter into space," the Square returned.

"Well, what you call space. But what do you, a Flatlander, really know about space? You only know two dimensions, length and width, but the space where I live has three: length, width, and height."

"Oh," the Square said, "we also speak of length and height, but that is the same to us as length and width. There are only two dimensions, after all."

"Not with us," the stranger said again. "I came out of a third direction which you cannot see, a dimension perpendicular to your two dimensions. I am the most perfect creature imaginable, an Over-Circle you might call me—Sphere is what we call it in Spaceland."

"You mean to tell me that a direction exists which is not north-south, not east-west, but perpendicular to that? Perpendicular to both directions at the same time?" the Square cried out. "Then please show me that direction."

"You cannot see it," the Sphere said. "To do that you would have to have an eye in your inside."

"An eye in my inside? In my stomach? But I wouldn't be able to see with that."

"I can look into your insides," the Sphere resumed. "Your inside is completely open to me. When I descended here I could see your entire house. I saw the occupants in their rooms; I could look into your closed closets." After a short pause he went on: "I have been observing you, dear Flatlander, for several days now. I saw a dream vision of Lineland in your brain. You were talking with the king of that country and you tried to convince him that you were able to look inside him. In the same way I, who am living outside of your flat plane, can look into your insides, because it is open to my view just as the inside of a Linelander was for you."

To make this clearer, the Sphere disappeared from the plane of Flatland and shortly thereafter gave a little shove in the Square's insides. He also took some objects out of the closed closets and deposited them in front of his host.

"Now I will descend once more into your space," the Sphere said. Again the Square saw a dot appear which turned into a small circle and slowly grew in size.

"I can see nothing of your descent," the Square said. "I can only see a Circle which keeps growing."

"You cannot see my ascent or descent," the Sphere

explained. "It occurs in a direction which is invisible to you. If one of your Circles were to visit Lineland, what would the king of Lineland see?"

"Well," the Square said, "first of all he would see a dot, which would change into a line. At least he would be able to determine that by means of the sound. The line would grow until it had reached its maximal value and then would diminish again."

"Exactly," the Sphere said. "The Linelander always sees the next cross-section of the Circle, and that is also how it is with you. You always see the succeeding cross-section of me and these cross-sections are circles."

"In other words," the Square resumed, "I can only understand this by means of an analogy?"

"Precisely," continued the Sphere. "By means of an analogy. And since we are talking of analogies, just suppose that a dot is moving northward, leaving a brightly lit trail behind it. What would you call that trail?"

"A straight line."

"And how many ends does a straight line have?"

"Two."

"Now just imagine," the Sphere continued, "that the straight line moves parallel to itself over an equally great distance to the east. What name would you give the figure which materializes?"

"A square."

"And how many sides does a square have? And how many vertices?"

"Four sides and four vertices."

"Fine. Now put your imagination to work," the Sphere continued, "and imagine that the square is moving parallel to itself in a direction which is inconceivable to you, entirely outside of Flatland; something then develops which we call a solid body. You cannot visualize it but you can calculate analogically how many vertices it has. Just remember: a single point consists of one dot. A line has two end-points, a square has four vertices. The series which results, 1, 2, 4, is apparently a geometric progression, and the next term therefore is . . . ?"

"Eight," the Square said.

"Right. You reasoned correctly. We call the body which materializes from the sideways movement of the square a *cube*. And a cube has eight vertices."

"And how many sides does the cube have?" the Square asked.

"That too," the Sphere replied, "we can discover by means of analogy. The border of something is always one dimension behind the thing itself. Accordingly, a line is bordered by dots, a plane figure by lines, and a solid body by . . ."

"Planes," the Square supplied.

"Right," the Sphere granted. "A point has no dimensions and no borders either. Thus, number of borders: zero. A line is bordered by two dots, a square by four lines. It is an arithmetic progression: 0, 2, 4, and therefore . . ."

"And therefore," the Square took over, "a cube is bordered by six squares."

"So you see you discovered the answer yourself, through reasoning," the Sphere summed up.

4 *To the Land of Three Dimensions*

Afterward, the Square could not have said how long the Sphere's visit lasted. Did it really happen, did the Sphere really lift him forcibly out of his plane, his Flatland, and take him to Sphereland, where he could view his world from the outside? He saw his own house as he had never seen it before. He saw the various rooms next to each other, his four pentagonal sons, both his hexagonal grandsons, his daughter, his wife, the servants. Outside, two policemen were strolling by. He saw the entire street, the theater, and in the distance a large polygonal building, the Council Room of the states of Flatland.

He came closer. A session was in progress. As was customary at a turn of the century, the states had gathered to protest against demonstrations from Spaceland. It was a matter of public record that at the beginning of the year 1 and also of the year 1001, a creature from beyond Flatland had paid a visit to the States Conference to argue for the existence of a third dimension. This time, too, a visit was expected. And it took place. The Sphere descended into the packed Council Chamber to carry out his assignment. Once every thousand years a delegate was sent out to try and convince the Flatlanders of the limitation of their world and of the existence of a three-dimensional world, and this time it was the Sphere's turn to undertake the mission.

His sudden appearance in the Council Chamber caused much confusion. The Supreme Circle ordered the guard to attack the stranger, but the latter escaped into an invisible dimension.

Thereupon the Council of Wise Circles resolved that no one was to hear about these amazing developments. Since the Circles were the only ones considered capable of keeping secrets, it was decided to kill all others who had been present. Only a few Chamber attendants were actually involved, and as a precautionary measure those selected for that post were less valuable Isosceles with exceptionally sharp vertices. The alerted guard arrested the unfortunates and carried out the verdict. There was only one other person present, the recording clerk—a Square who was recognized with great shock by our Square as his own brother. The poor man, who could only be accused of having had to be present at this session because of his official function, was not sentenced to death but only to solitary confinement for life.

After completing his trip, the Sphere returned to his guest. He showed him a cube, but since the Square was not used to seeing perspective he mistook this body for an "Irregular." Still, on the basis of what he could see and by applying the analogy, the Square succeeded in getting a pretty good image of the world of three dimensions. Finally he

A Cube (A). *The Square mistook it for an Irregular* (B).

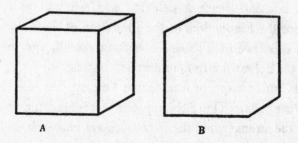

A   B

turned to the Sphere and said: "From the land of two dimensions one can see the inside of Lineland's inhabitants. You have taken me with you to the land of three dimensions, from which I can see the insides of my compatriots. Now I should like to go to the land of four dimensions, from which I can take a look at your insides."

This was too much for the Sphere. He considered himself the most perfect creature imaginable, and now this puny, two-dimensional creature wanted to look all the way into *him*. Besides, in his opinion a land of four dimensions was impossible. When the Square told him that the analogy could be continued and that a four-dimensional body could therefore be predicted, an Over-Cube which according to the progression: 1, 2, 4, 8, 16 must have sixteen vertices and according to the progression: 0, 2, 4, 6, 8, eight side-cubes, and when he asked whether no fourth-dimensional visitor had ever come down into the land of three dimensions—then it all became too much for the Sphere. He did admit there were rumors that some people had visits from a very strange creature out of another world, but no one had ever taken this seriously. It was generally ascribed to hallucinations arising in the brains of sick individuals, the result of their "confused angularity"!

When the Square persisted, the Sphere rudely shoved him back in his study, where he gradually returned to his senses from the daze brought on by shock—not certain whether he really made the trip to the land of three dimensions or only dreamed it.

He went to his bedroom and fell into a deep sleep. And

here he dreamed again. The Sphere took him to the country of zero dimensions, consisting of only one single point who was humming along self-sufficiently and happily, thinking he was all there was in the world—and actually he *was* everything in it, because his world did not have one single dimension.

5   *Dishonored*

The following day, New Year's Day, our Square found himself facing the new millennium with joy and anticipation. This was to be the millennium of enlightenment. There would be epoch-making new ideas, starting with a better insight into the existence of a world of three dimensions. He felt himself called to preach the new doctrine. The concept "Upward, not northward," engraved in his mind, would be his guideline.

With whom should he start? With his wife? At the very moment he was considering this, he heard the voice of a town crier on the street announcing the Council's decision that anyone trying to poison the mind of the public with claims of getting reports from another world would be imprisoned or killed.

This was not to be taken lightly, but the Square was self-confident and strong. He would not make any loose statements, but he could give a scientific demonstration, an argumentation, and that would change things considerably.

Nevertheless, it might be wiser not to start with his wife,

much less with one of his four sons. Not one of them had more than an average aptitude for mathematics, and besides, he was not sure that their filial love would outweigh their sense of obligation to report their father to the prefect—their father being a mere Square and they themselves Pentagons!

He thought it best therefore first to test his grandson, who had undeniably great mathematical aptitude and had made such intelligent comments. He had the boy come to him and tried to explain that the third power of three could indeed have a geometric meaning. He told how a square can originate from a line, and how, by moving this square upward and not northward . . . But already he found his words beginning to lack conviction, for when he wanted to make himself clear by moving a square, he found himself pushing the square in an arbitrary direction and repeating, "Upward and not northward"; but of course he could not move the thing upward.

At that moment the town crier came again within earshot, threatening all who propagated revelations from another world with the most terrible punishments. The boy also heard this and understood it all too well. He became terrified, and bursting into tears he said that he had not meant to imply anything and that something like this could have no geometric significance at all.

Any further attempts to convince the boy of the correctness of the concept he had suggested earlier were of no avail. The boy was afraid that his grandfather had it in for him and, terribly frightened, he ran out the door.

The Square thought it best to keep his view to himself, but it was not easy, because anyone who has acquired such an insight naturally burns with the desire to make it known to others. And so it happened that at a gathering of the Natural Sciences Association—and, worse, at the very home of the prefect where one of the members was giving a lecture on Providence's foresight in limiting the number of possible existing dimensions to two, thereby leaving it up to Providence alone to view the innermost parts of things—it was at this gathering that he raised his voice and in a fiery dissertation argued the existence of the third dimension. He told about the Sphere's visit and about his own trip to Spaceland.

Of course he was arrested, and inasmuch as he could not meet the only requirement asked of him, to point out the direction of "upward but not northward," he was sentenced to life imprisonment.

In solitary confinement in a dungeon he was allowed to be visited once a week by his brother, who, as we know, had also been sentenced there for life. With him the Square spoke regularly of Spaceland, but without any success. His brother, who had actually been present during the Sphere's visit to the Council Chamber, did not want to accept his theory.

Lacking any followers, the Square began to write his memoirs in the hope that these might at a later day come into the hands of more enlightened minds who would be able to see the truth of the existence of a space of three dimensions.

And these are the memoirs which under the title of *Flatland* have been published in an enlightened world, now fully capable of understanding their meaning and significance.

SPHERELAND *A Fantasy About Curved Spaces and an Expanding Universe*

by A HEXAGON

PART I The Straight World

1 *Changed Times*

More than seventy years have passed since my grandfather, the famous Square, published his ideas about other worlds. I consider it my duty to show how greatly these ideas have now changed. And I feel myself called upon to do this because a slight but constant feeling of self-reproach has been bothering me and I have not been able to reason it away completely. Wasn't it I who, against my own better judgment, deserted my grandfather? At the time of the discussion in his room when he tried to explain to me the possibility of the existence of a third dimension, I put him off and acted as if I thought he was talking nonsense—and all the time I knew perfectly well that he was right. Even though I try to soften this by telling myself that I was only a child and therefore not responsible for my actions, and that I was afraid of the consequences, self-reproach has stayed with me through the years.

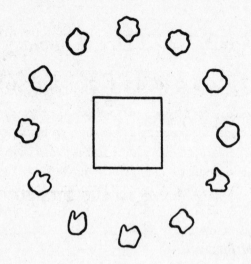

My grandfather's statue in a circle of twelve trees.

My grandfather's honor has now been restored. Unfortunately he did not live to see this, but his descendants have done everything in their power to right the wrong that was done. A bigger-than-life-sized monument has been set up for him on the market square of the town where he was born. It is surrounded by a circle of twelve trees and this circle represents the highest wisdom.

Each year a delegation including several professors of mathematics and physics, the mayor of the town, and notable figures in the field of science and politics pays an official visit to the monument. A speech is made recalling how backward society was when it believed it could restrict or curtail science.

But great changes have occurred in other areas too. The

backwardness of women has disappeared almost entirely. The notion that a female is a stupid creature because of her small brain area has turned out to be false. Some women now even study and get degrees from a university. Man no longer has a monopoly over science, even though nature did create woman first of all for marriage. It is she who by virtue of her great gifts of love and devotion has been destined to raise children and to dedicate herself to homemaking. This takes up such a large part of her mental life there is usually no room left in her intellect for study of science. The really important inventions and discoveries will undoubtedly continue to be made by men!

Even woman's temper tantrums, once so greatly feared because they could cause such disasters, are mostly a thing of the past. Nature has endowed woman with the gift of self-control. This was not understood formerly and girls' education helped make them think they were totally lacking in it. Therefore, no feeling of responsibility was developed and a woman remained convinced that she could indulge in rages with impunity. After all, she was only a woman! Now she knows, however, that like the male offender she will have to answer for her behavior before a judge.

Another very important factor enters here. As a rule, the judge will order an examination of a female defendant's mental capacity in order to determine whether she is responsible for her actions. With us such an examination simply consists of checking the external shape of the defendant, and a woman is mortally afraid of even the smallest deviation being found in her "line." It would be a blow to her pride.

She much prefers to be held fully responsible, even though she has to bear the consequences.

The danger that woman presented to society largely disappeared with the innovation of the shoe for women. In public, every woman wears a little shoe on her posterior. You can even collide with it at top speed without seriously hurting yourself. The law that a woman must constantly move her endpoint back and forth has been repealed, there no longer being any need for it, and a woman no longer needs to sound constant warning cries when out in public. The traditional way of moving through the streets did remain in vogue for a long time among women of the upper classes. They were proud of the graceful, rhythmic movements of that part of their anatomy, after all! But now even this has disappeared. Aristocratic ladies became such objects of ridicule for boys in the street that they began to feel silly; moreover, it happened more than once that their rhythmic movements made them lose their shoe. Then the young urchins would make fun of them, yelling "Cinderella, Cinderella!"

To understand this, you must know that we have an old fairy tale about "Cinderella." She was a beautiful young girl, very straight and slim of line. And she had two stepsisters who were also very beautiful, but not as beautiful as Cinderella. They were, in fact, considerably heavier of line. Naturally this made them very jealous and so they forced her to do all the dirty work. She had to fix the meals, scrub the house, and clean out the fireplace. And because she was always brushing and blowing in the ashes, she could never get to look like a lady.

As fate would have it, an official court ball was to be given in honor of the crown prince, who had just reached marriageable age. The most beautiful girl in the world was to be chosen for him, and all young ladies who thought they might qualify for princess were invited to come to the palace at nine o'clock. All young men between eighteen and twenty-two were also invited. Their particular qualification was that they should have at least six sides, but control over this was not very strict.

Cinderella would have liked very much to go to the ball, but her wicked sisters laughed at her and said that she could not go without a dancing shoe. It was the custom that ladies at a ball were to carry a shoe at all times, since serious accidents might otherwise occur with the rhythmic back-and-forth movements in dancing. And so Cinderella stayed at home. She sat by the fireplace and dreamed of the prince, a handsome young twenty-four-sided polygon who had been the topic of everyone's conversation of late.

While she sat there brooding, a neighbor came in and said: "Dear child, I will give you a dancing shoe such as no one else possesses—lovely, elegant, and transparent. But first you will have to fix yourself up." When Cinderella had washed herself thoroughly she put the little shoe on. It fitted perfectly. She thanked the kind woman and hurried off to the palace. The neighbor called out after her that she would have to take care to be home before midnight, because the little dancing shoe was made of some chemical material which remained hard and solid for only three hours, after which it would start turning into a jellylike substance. And she would

look ridiculous with such a gooey mess stuck on her body.
Actually this was not true, because the shoe was made of
beautiful glass that could withstand centuries. But her bene-
factress was shrewd. She knew a good deal about people and
understood that the prince, who would of course fall in love
with Cinderella, would receive a shock when his lovely dance-
nymph suddenly disappeared.

And that is exactly what happened. When Cinderella
made her entrance, the festivities were already in full swing.
Everyone looked up and the prince immediately hurried over
to her. In fact, she made a spectacular impression on all the
young men present. Her glass slipper set off her lovely body
line very nicely. The prince danced only with her until the
clocks struck twelve. Cinderella started up, tore herself free,
and ran off. The palace doors were closed but she escaped
through the ventilator. The prince, who had run after her,
was much impressed by the fact that she could go through
such a narrow opening. Of course he could not follow her,
but what was that? He had seen the small, delicate shoe which
she had lost during her escape through the narrow hole. He
took it with him and swore that he would marry the owner
of that shoe and no one else.

The next day he went through town, preceded by heralds
who informed the people of the prince's intentions. Every
young lady could report in person to try on the glass shoe.
Many had to admit with disappointment that the shoe was
amazingly delicate and small. Cinderella's sisters tried it on,
but of course without success. Then Cinderella asked if she
might try it. Her sisters laughed at her and said scornfully:

"You, Cinderella? You mean that you want to marry a prince?" They hadn't recognized her the night before, but they had jealously seen how the prince danced constantly with the same beautiful young lady and would not pay any attention to them. That this young lady could have been Cinderella did not occur to them.

But Cinderella was also allowed to try the shoe on, and to everyone's total surprise the little shoe fitted as if it had been made to order. And now the prince recognized his beloved and brought her triumphantly to the palace. They were married a short time later and lived long and happily, with many children—nice twenty-five-sided polygons and lovely, slim girls.

2   Easing of Class Consciousness

Though the class consciousness which dominated our society for centuries has not disappeared altogether, it has lost much of its sharpness. When it was realized that the female does possess intelligence and that her backwardness was primarily due to insufficient training, it also began to be understood that the size of the brain angle is not an absolute measure of intelligence. It is true that we will never find a scientist with a very short base among the Isosceles, not even a boy with any aptitude for advanced education, but it has happened repeatedly that an Equilateral has become a professor!

The increased freedom of association between the two

sexes has also done much to lessen class differences. Already little difference is to be noted among the girls of different classes, and we know that even in the old days it was not unusual for a Circle to choose for his wife the first girl who happened to charm him. The many sides he possessed made him unconcerned. What if his descendants had a few sides less than he did?

This indifference gradually spread to every group. First it affected the Polygons with many sides, later also the Hexagons, the Pentagons, and even the Squares and the Equilaterals. An Equilateral who as the result of an ill-chosen marriage had Isosceles children was no longer ashamed of it and did not reject his sons, but kept them at home with him, bringing them up as lovingly as if they had been born pure Squares.

On the other hand, a sharp decline in ambition was noted among the Isosceles. I knew an Isosceles with a vertex of 59½°. He had made a very good marriage and his sons were all pure Equilaterals. He would not think of presenting them to the Caste Council for testing, however, and when the form of his offspring was investigated at his home, he threatened to "find" the officials again if they dared to pronounce his sons pure Equilaterals. But the officials could not be put off, with the result that the boys were taken away to be raised— as prescribed by law — far from their parents in the family of a childless Equilateral. The father then called the entire neighborhood together and the boys themselves also resisted bitterly. The police had to enter into it at last before order could be restored. Still, the boys would not adapt themselves to their new environment. They said that father-love was a greater virtue than equilateralness and they continued to visit their parents every Sunday, though their foster parents had forbidden it. In desperation the latter asked the authorities to release them from their task of bringing those rebellious youngsters up properly.

The Caste Council thought it advisable to return the boys to their parents, thinking to use this as a horrifying example, but the consequences were quite different. When an Equilateral was born into another Isosceles family, the father announced it publicly, adding that he would take the boy home and bring him up himself. The Council was very upset by this but thought that it could solve the problem by offering to enter the father as an Equilateral in the records of the civil registry—though his vertex was just short of 60°. But the man refused this offer, saying he preferred to go through life as an Isosceles of stature and rank rather than an unfit Equilateral who would be given the cold shoulder by everyone.

Opposition now began to increase by leaps and bounds. Slogans even started to make the rounds of the Isosceles, proclaiming: "Remain yourself. Do not strive for Equilateralness!" and "Do not marry a woman with a pedigree or you will go down into the ranks of the Equilaterals!"

All this was much worse than a rebellion, because it did not constitute a protest against the higher ranks but the sowing of a widespread contempt. It assumed serious proportions. In the case of an Equilateral, it is easy to see on what side the eye is located though the three angles are of equal size. But the rude practice now developed of addressing an

Equilateral at one of his other vertices and when there was no answer to act as if a mistake had been made, exclaiming apologetically: "Please excuse me, but in your case it's so hard to tell front from back!"

As a result of this systematic disrespect, a feeling of self-respect developed among the lesser ranks. Add to this the expansion of education and it is clear that the self-sufficiency of the Isosceles ranks was bound to develop further. In addition, the medical sciences have made great progress and operations designed to increase the number of sides have become much less dangerous and expensive. Where formerly the success of such an operation was the exception, failure of it now was called an unfortunate accident. Recently an Isosceles had his three sons undergo simultaneous operations. After a few weeks the boys returned home from the institute. One had become a Hexagon, the second a Nonagon, and the third a Dodecagon—all three beautifully regular. Instead of the entire family being highly respectful toward the Dodecagon, not only the little brothers but even the triangular father felt free to tease him, calling him "Do Decay," "Little Twelver," or "Pint-sized Dozen." An offer from the Caste Council to have the young Dodecagon enter the ranks of nobility was rejected. The father thought it ridiculous and the son responded in the same spirit. From that time on many nicknames were invented for the boy and he was sometimes called "Dodeca Count," "Baron of Twelve," or "Duke Dozen"!

All this must be seen as a reaction against the exaggerated respect for the upper ranks so prevalent in earlier times. As with all reactions, this one was only temporary. Yet something of it will remain. The exaggerated reverence for a Circle which the old guard would like to see return now belongs to the past forever. Today more attention is paid to individuals, and an Isosceles with great gifts is valued more highly than a lazy Circle who is living off his ancestral reputation.

It is odd that even more attention is paid today to someone's background and origin than in the old days when it was rare for anyone to enter a much higher rank all of a sudden. A twenty-four-sided polygon created out of an Equilateral after a successful operation is generally not considered to have the same pedigreed stamp as a Hexagon who has steadily climbed the ladder of successive development. The linealogy registers always indicate exactly how the person in question has acquired his many-sidedness. I don't believe that appreciation of good stock will diminish in the future. Too much depends on it and this is sensed even by the less-developed ones who are usually the loudest opponents of class differences. Generally, however, the schools of the upper ranks drill the realization into the youths that nobility carries obligations with it, and they are constantly reminded of the wise saying of an old philosopher: "Whatever you inherit from your fathers you will have to earn to possess it fully."

And so this quiet revolution had important consequences —not only politically, but scientifically. The new conditions had a refreshing effect. Science dared to cast off the old. New ideas could be considered, and a scientist who proclaimed the existence of the third dimension was no longer regarded as a criminal or a dangerous lunatic. It is sometimes said that in the days of old anything new was accepted only when officially proclaimed by the Supreme Circle, and there is some truth in it. But today it's no longer a question of *who* was the first to have a new insight. A square can also have useful ideas—as did my grandfather.

3   Explorers' Trips

Another reason for the general change in attitude is the broadening of insights as the result of expeditionary journeys. Only a short time ago it was thought that nearly all of the world had been discovered and was now known, and that furthermore only those areas which had actually been visited were of any importance to mankind. That there were still some unvisited areas goes without saying, because the world is infinitely big.

Everyone was convinced that all inhabited countries were inhabited by the same creatures that were living with us, and that at most only the levels of development and culture might differ. It was thought that the most highly developed beings were to be found in our society and that more highly developed creatures were unthinkable—in this day and age at least. Of course an improved and ennobled race might inhabit the world in the future. Perhaps a society would come into being in which all men would be Circles and all women very graceful, slim lines, equally noble of mind and spirit. Crime would no longer be known. Everyone would be splendid and good, carrying out his assigned tasks with joy and dedication in this lovely world of the future. There would be much time for recreation which in turn would be used to improve the body; everyone, moreover, would voluntarily strengthen his mind in his spare time with spiritual food. Any instance of crime or, rather, "less-refined behavior" would be viewed as a symptom of a disease to be treated and healed in the clinics.

This then was the dream of an ideal state in the far-distant future as the end product of the propitious evolution of our basically so noble race. But that we ourselves, in the present-day world, were the most highly developed, of that everyone was convinced—except perhaps for one or two philosophers. They dared to argue the possibility that in some far-off country other, much more highly developed creatures might live who would consider us an under-developed race, a sort of domesticated animal perhaps. But this could not be verified. Wherever our explorers went, they always encountered creatures like us, usually at a level of development much lower than ours.

For example, a country was once discovered where the vertical angle of the men is never more than 10°. At first it was thought that this was a degenerate race, but upon closer investigation some deliberate scheme appeared to be involved, since all boys with a larger vertex were simply dispatched. Further inquiry revealed that more developed creatures had lived there at one time—Equilaterals, Squares, and even Polygons—but that a revolt had put an end to them. In a battle which must have ended in an enormous bloodbath, the very pointed Isosceles retained the upper hand. They did not stop fighting until all higher-ranking creatures had been killed. Inspection showed that the only survivors were Isosceles, most of them with a very small vertical angle. By a majority vote it was then resolved to establish an upper limit of 10°. Those having a vertex larger than this were ruthlessly killed off by the common army and with that a new order was established.

At first, life in this state was gay and happy. It was a powerful group of people and they emerged victorious from every war. But it became evident after a while that something had impaired public welfare. It is true that the Isosceles with a vertical angle of 10° were not completely devoid of intelligence, but practitioners of the arts and sciences were almost entirely missing. Intellectual life had faded out and the entire society was doomed to decline. In the long run, competition with other countries which had not experienced this "equalization" could not be kept up. While the country was not doomed to total destruction, it had to take a back seat, and will certainly not be able to take its former place culturally among the other nations for centuries to come.

In another country, not far from this one, the inhabitants had only been interested in intellectual development. It is true that physical education was not altogether neglected and that all residents did participate in sports, but there was no army. It was felt that there was no need for one, and as the result of a fine harmony between body and mind, evolution had made such strides that scarcely any Isosceles were left; those who were, had a vertical angle of more than 50°. At the same time, the country was overrun with

Polygons and Circles. Life in this land was nothing short of paradise, in fact—until a cruel disenchantment set in. A neighboring state, which maintained a powerful army but was less successful in carrying on peaceful relations with other nations, grew envious of all the riches piling up so close to, and yet outside, her borders. Following a theory of her own concoction, namely, that all the riches of the earth belong to the most powerful, she made a surprise attack on her peace-loving neighbors, killing most of them and taking possession of the country's resources while enslaving the remaining inhabitants.

Farther to the south, where the force of gravity is stronger and the atmosphere more oppressive, entire tribes were discovered living at a very low level of development in dense forests. No Equilaterals or more highly developed creatures are to be found there. They are all Isosceles and, what is more, they all have a very small vertex. In certain tribes the head angle is not more than 30°, and in others it is even smaller. Discovery was also made in these forests of pygmies, dwarfs, who never grow taller than children do in our land. This discovery surprised the civilized world very much. Dwarfs often do appear in fairy tales, but no one had ever thought that such creatures really existed.

In the midst of dense, dark forests a tribe of Irregulars was discovered. As was understood right away, these crea-tures had to be inferior intellectually—a tribe of born criminals, in fact. Fortunately for the civilized world, most of them were unable to go beyond their own territory. They were too big and clumsy to move in between the trees and so were forced to live in the open spaces which remained in the forests and which they kept cleared themselves. For the sake of harmonious coexistence, the civilized people then decided to band together and exterminate these tribes, which were a travesty of society. Nearly all nations cooperated voluntarily in the cleanup—all the more so since resources discovered in the forests of the Irregulars were known to be of particular value to civilized countries. As a matter of fact, slight differences of opinion about this have even resulted in wars between civilized nations.

It would take me too far afield to elaborate on all the strange peoples discovered in our world. I just want to tell you briefly about the Amazons. This was a group of people thought to consist entirely of females, because no one had ever seen any men there. Many theories had been worked out to explain how procreation of these creatures was possible, and the experts were still heatedly debating them, when all theories were suddenly found to be wrong. Men *did* exist in the Amazon state, but they were living in confinement. They were very few in number and consisted of Squares and Polygons. The boys who were born as Triangles were immediately destroyed, even the Equilaterals, together with as many of the remainder as the society could do without. The rest of the men were kept together in a camp where, in addition to maintaining the race, they were used for household tasks. A few female custodians could easily control them—at the least sign of resistance they were promptly perforated.

History records that a rebellion once broke out among the men. They overcame their female wardens and tried to escape. A whole army of Amazons pursued them, and since the sharpest-angled men were only Squares, and others having even blunter angles, they were unable to defend themselves. As a result, they were mercilessly destroyed by the onrushing horde of females who quite literally riddled them.

The Amazon nation was now threatened with ruin unless measures were taken to import new males. These had to be obtained from neighboring states, which did not occur without battles. But although the Amazons suffered many casualties, these well-trained fighting females managed to maintain the upper hand every time and they carried off a large number of male prisoners.

If this had happened only once it would not be so bad, but the first successful attempt had provided the Amazons with a taste for war. It seemed much easier to steal adult males than to bring them up from children. Consequently, all boys born among the Amazons were killed and only girls allowed to live. The latter received training as soldiers, uninterrupted by any distraction such as the presence of boys. The supply of men was simply replenished from time to time with another foray.

It goes without saying that the neighboring states were none too happy about this, especially since successful raids had made the Amazons very reckless. The Amazons robbed and looted everything they fancied and made several nations tributaries to them. One day the tributaries rose in revolt and organized a large army which marched against the Amazons. In a bloody battle, in which both sides suffered many casualties, the women were victorious. Understandably, the revenge conceived by Bellaforta, queen of the Amazons, was horrible. All the boys in the conquered countries were put to death without mercy and the girls carried off as prisoners. The very young girls were trained for service in the Amazon army; the older ones, who already appeared attached to their family and home country, were dispatched.

Nations which until then had not been disturbed by the Amazons now became frightened. A large convention of heads of state was called together and the serious situation discussed in detail. A proposal to launch a female army against the Amazons was rejected since its training would take too long. But it was clear that an able, well-trained army was called for, one made up of fine soldiers with very sharp vertices.

The campaign, in which I myself took part as a very young lieutenant, will always be recorded with glory in the annals of Flatland. The general, Prince Armatus, was a decagonal, a man of exceptional strategic insight. With his army he waylaid the Amazons at the edge of a big forest by placing a tightly closed phalanx directly in front of the trees, so closely together that the forest could not even be detected from the outside. At the last moment when the Amazons came charging up in a wild rush, the soldiers slipped behind the trees. The wildly flying women saw the danger too late. With their sharp points they bored themselves into the tree trunks. Many females broke right off in the middle and others remained rooted in the wood, becoming an easy prey for the infantrymen, who attacked them from the side.

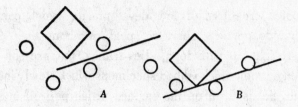

The queen of the Amazons attacked by Quadratus
(A) and trapped between three trees (B).

Queen Bellaforta and her staff, who were right behind the onrushing troops, took to their heels. They were overtaken and driven sideways in the direction of a cluster of trees. There they thought they could cleverly slip away between the tree trunks, but instead they were attacked from the side by carefully hidden troops who had been lying in wait for them. A sergeant named Quadratus pushed the queen sideways at precisely the right moment so that she became stuck between three trees, unable to move frontward or back. They carried her off in triumph and put her on public display at fairs and circuses in many different countries. The brave sergeant, who immediately became a public hero for his great presence of mind, was changed into a Dodecagon at the state's expense and made a nobleman.

4   The Trees, the Wildlife, and the Sea

Although we have talked repeatedly about trees and forests, you still may not know what the trees in Flatland look like. As is true in your land of three dimensions, they are rooted creatures of low development, which can only feel, but do not see, hear, or speak. With us they do not grow straight up with long, slim trunks, but are somewhat rounded or sometimes ribbed specimens which slowly increase in girth. Seeds form on the surface which pop off and take root wherever they land, and from them new plants develop in turn. When they have reached a certain age, which in the case of some varieties can be more than a hundred years, they die and dissolve in the air.

It can happen that the trees of a forest become so thick that they grow into each other, becoming an "impenetrable" forest in fact. But even before it has reached that point, a forest can be so dense that full-grown individuals are no longer able to enter it; children can still squeeze in between the trunks, but there is great danger they will get lost in the woods. They lack the orientation capacity of adults, who can pretty well sense gravity's direction (southern) and who can also determine from the color on the trees where north is. All the trees are somewhat green of shade and have a fresher color on the north side—this being the side which receives more rain.

Many different types of trees have been discovered. In the southern areas they are more moist and green and there they grow more quickly. The atmosphere is damper there and this sometimes makes walking through the woods a far from pleasant experience for the Northerner. It is quite oppressive and depressing, and it happens often that a traveler in a dense forest is suddenly gripped by great fear—

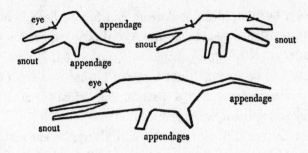

Assorted dragons.

not without reason, since the woods are inhabited by animals which are dangerous to the traveler.

Since these animals cannot be compared with any animal type in your Spaceland, we'd do best to call them "dragons," though they differ greatly in form and must therefore obviously belong to different species. If you should like to know more about this, I shall be happy to recommend several zoology textbooks to you in which these odd animals are described. Generally the dragons have a rough surface with ribs, or scales, or whatever you wish to call them. They further have a few hollows or openings, snouts, through which they maneuver other creatures inside, either other dragons or more highly developed creatures. The snout then snaps shut and the prey is absorbed inside the body.

That life in the midst of those monsters is not entirely without danger is evident from the fact that a recent scientific expedition into the jungles for a systematic study of flora and fauna never returned. Woodcutters working at the edge of the forest heard loud shouts and cries for help far away, but they did not dare to run in and help, sending word instead to a nearby town. A military expedition was sent out at once and could still make out where the calamity had occurred. Right between bushes and trees they found some remains and a part of the scientific notations made by members of the expedition.

Even though different peoples will always be encountered when one is traveling east or west, the inhabitants of our world are interested more in what the south has to offer, because when you travel to the east or to the west, the conditions of climate, gravity, and air pressure all remain the same, while the gravitational pull and air pressure increase toward the south. Besides, the temperature is higher and plant growth more luxurious there.

It soon became apparent that not everyone was fit enough to undertake the trip to the far south. Many became the victims of a tropical disease called "dengue," of which a patient can die within a few hours. The moment the first symptoms appear, the sick person has to leave for more temperate zones. Usually he is no longer able to do this without any help. But since the sick person is unable to travel unaided by the time the disease is evident, those who are traveling alone at that time are as good as done for. Even large-scale expeditions have succumbed to this tropical disease, however, which is quite understandable. To undertake such a trip in that unhealthy, sultry climate requires much pluck and perseverance. The men who set themselves this task have only one goal in mind: to push on. They will persevere and penetrate the dangerous country as far as possible. When one of them falls ill, the others cannot decide to turn back as long as they are still healthy themselves. And so they move on, carrying the sick man with them or leaving him behind in the care of one or two others. The result usually is that the entire expedition perishes.

As a result of heavy exercising and firm discipline, together with strict medical examinations, groups of men were eventually formed who could withstand the climate's bad effects, and with the help of such groups the southern regions were finally carefully explored. It was discovered that the vegetation became more luxuriant toward the south and that the humidity increased sharply. Trees grow rapidly there and their surface is usually covered with moss or mold which ultimately causes them to suffocate. It is common knowledge that dead plants and dead animals are quickly dissolved in the air. This happens even more rapidly in hot regions than in the more temperate zones. Nevertheless, the decaying vegetation in the jungles has a bad effect on respiration, with the result that the concentration powers of the expedition members, who are already feeling the effects of the oppressive atmosphere, diminish even more. They feel dizzy, develop a ringing in their ears and can no longer think clearly. An expedition leader will have to decide the right moment to turn back. And of course science is benefited more by a successful expedition which does return and is capable of recording its experiences than by one which pushes on but fails to return.

In spite of the many hazardous and difficult conditions an expedition would have to face in these forests, the scientific explorations continued. In this way many scientific data were collected which indicate that the forests are becoming increasingly impenetrable, not only because of the ever-increasing growth of trees, but also due to the constantly spreading overgrowth on the trees. Finally the explorers encountered forests that were absolutely impenetrable as a result of creepers and trailers growing in between the trees and blocking every passageway.

Since people still wanted to know how far south one could travel and whether an open area might yet appear beyond the forests, an expedition was launched equipped not only with scientific investigators but also with a small army of trail hackers. These were very sharp Isosceles especially trained for work in hot areas. It became a famous expeditionary tour which lasted a long time. Doubts began to arise about its return and it was feared that the entire group had either succumbed to tropical diseases or had been devoured by dangerous and still unknown dragons, when the report spread like wildfire through the civilized world that some members of the expedition had returned. They related their experiences of penetrating the dense forests at last and coming upon a sea, the world sea in which rainwater was collected. They had seen marvelous creatures at the upper line of that sea, animals which, unable to lift themselves up from the ocean's upper shoreline, were easy to catch and provided the expedition with delicious food.

The returned men were honored and feted. The scientific leaders gave public lectures and talks to learned associations. They were regarded as the discoverers of the world's outer limits, for who would dare start an investigation into what is still to be discovered below the upper line of the world ocean? But, instead of putting an end to exploratory journeys, this expedition was the beginning of increasingly daring attempts to explore the world.

In less than a year another expedition set out, equipped with the necessary means to submerge below the ocean's upper line in hermetically sealed little boats. The men returned with descriptions of animals seen beneath the ocean mirror and they told of fights between these creatures. It might be simplest to call all these animals living under the water simply "fish," but you must not think that the sea monsters differ from each other only in degree. On the contrary, they vary greatly in form and size. Some are almost round, others are lancet-shaped, and still others are irregular. Some seem to have long tentacles, and such a creature once got a stranglehold on the expedition's little research boat. The crew had already abandoned all hope when the beast suddenly loosened its hold—it had wounded itself on one of the boat's sharp corners. The waters turned red with its blood, but whether the creature perished could not be determined, for it rapidly disappeared from sight.

New expeditions were equipped to descend to the bottom of the ocean. Here fantastic, luxurious vegetation was found in which a man could easily become hopelessly entangled. It was even possible to drill into the bottom of the ocean and there the ground was found to consist of a hard substance which we call "stone" or "rock."

In other words, the "rock-bottom limit" of the world had

now really been discovered and only philosophers could still discourse about the thickness or the thinness of the rock layer, whether it is an infinite mass or conceals something else again, e.g., an open space, another world, or whatever. For the time being, science was unable to answer that question.

5 *The Trip Around the World*

Coincidence can sometimes play an unexpected role in history. It is obvious that a scientific expedition, equipped with great effort and expense and entailing great danger for its members, does not always produce proportionate results— but that a quite simple exploratory trip turns out to be of exceptional importance to science is surprising indeed. Yet this is what happened with a trip which had been undertaken, not by a scientific expedition of experts, but by two vacationers as the result of a bet.

In a certain town there was a society called the "Club of Squares," because only Squares were eligible to become members. The club included two very widely traveled individuals who were renowned braggarts. One time, when the conversation had once again turned to the subject of strange countries and peoples, their friends set the two against each other, egging each of them on to establish once and for all which one of the two had done the most traveling. A committee was appointed as referee but it was soon faced with a grave problem. Apparently one of the two gentlemen, Mr. Orientalis, had visited more countries and come to know more people, but the other, Mr. Occidentalis, had been a greater distance away from home. One of these two must be granted the highest honor, and after long deliberations the committee felt Mr. Occidentalis was the winner. Mr. Orientalis then cried out that he knew how to travel much farther than his rival had ever done. The result was a bet. Both men would go on a trip—Orientalis to the east, Occidentalist to the west. Their respective reports would determine to whom the highest honor should go.

Orientalis had decided to take a long vacation and make an extended journey of several weeks, but when he learned that his opponent was apparently making preparations for a genuine explorer's journey, he acquired equally elaborate equipment. Then he arranged his affairs in such a way that he would be able to stay away as long as a year if need be.

When the appointed day arrived, the members of the club gathered together, the chairman gave a speech impressing on both men the seriousness of their undertaking, and after that the world travelers took off in opposite directions.

The various possibilities of this venture formed the main topic of conversation at the club and the members would often speculate on where the two men might be, whether they were still pushing on or might already be on their way home. Many bets were made, which indicated that neither was a favored entry, their chances being put equally high.

When a year had passed, however, without the return of either one, the members began to worry. The two were well known for their stubbornness. It was therefore possible that each had pushed farther and farther until he landed among wild tribes who were now holding him prisoner or had perhaps killed him. It was seen as a grave oversight that no deadline had been set within which they must complete their journey, for even if one now did return he could not be acclaimed the winner since there was no way of knowing whether the other was still alive and might yet report back to the club. The chairman then proposed that the first one to return be named the winner; if the other were to come back later from a longer journey, he could be given an extra celebration. This proposal was well received, but days went by, then weeks and even months, until finally a second year had passed. Concern increased; the case was now only discussed in whispers as if the talk concerned dead men. Some guilt was felt about their deaths, inasmuch as they had been needled into risking this senseless venture.

A third year went by and no one dared breathe another word about the painful subject. But a few months later the two travelers, supposedly lost and long since gone forever, marched into the club side by side and sat down among their friends as if nothing had happened. Mr. Orientalis announced that his friend had won the bet and for the rest of the day they did not say another word.

It was not until much later that they gave a detailed report on their experiences, which were so unusual that the entire scientific world became involved. What actually did happen? Orientalis had headed east. Like his opponent, he had carried an accurate gravity meter to help keep him on a straight easterly course. Whenever the instrument indicated a slightly smaller value, it meant that he had veered a little more north; if the gravity pull increased, it was a sign that he had wandered off toward the south.

For many months he roamed from city to city, country to country. At first the countries were familiar and civilized, but later on he entered areas of more primitive people who regarded the world traveler with awestruck amazement. Since less-developed people, who naturally see a Square only seldom, usually have a boundless respect for more highly developed creatures, whom they view as skilled medicine men, he could travel about easily. Everywhere he went he was honored as the infallible medicine man. Often he would be called to the bedside of the sick, and if a patient recovered he received the credit. When there was no cure, this was ascribed to evil spirits.

And so he pushed on from village to village until his journey came to a sudden end. His movement had been impaired as the result of a slight accident and he was very lovingly taken in by the nearest village, which regarded him as a magician sent from heaven.

Here he continued to live for months. He was given fine food and drink, but gradually an irresistible homesickness came over him. Was he to waste the rest of his life in exile? With pointed Triangles and even Irregulars? Although he was revered as a god here, that honor, awarded to him by primitive people, meant very little. If only there were someone nearby with his own background and development, he would be able to stick it out here; but he was all alone in the midst of a primitive tribe.

Often he would try to keep up his spirits by arguing

that he was living happily here since he could bring happiness to such an underdeveloped people who really ought to be regarded as children, but that did not help much; his black moods increased with time.

He began to recover and soon he could move about again; only short steps near his house at first, but after a while he was his old self again and sometimes made full-day trips to keep in practice. He was wise enough to wait until he was completely healthy and no longer got tired walking. Then the time had come to leave, he thought, and he wanted to go home. The extended rest period had killed all his desire to journey. His opponent was bound to have returned long since and would now be telling colorful stories about the many countries and peoples he had seen. He could imagine the club members sitting around the storyteller and listening spellbound to his tales. But what would they think of *him?* That he had gone much, much farther and ought therefore to be named the real winner? Or would they think he was gone for good?

These speculations made him feel low, not so much for himself as for his competitor, who would lose the honor due the latter. But soon he, Orientalis, would be able to travel again and he'd return as soon as possible, enter the club, march straight to his friend, and greet him as the winner.

For a brief moment the thought occurred to him that he might proclaim himself the winner. He could invent some stories about extraordinary people and unique countries where neither a Square nor a Polygon had ever set foot, but his honesty would not let him. He would have to be a hardened liar to acknowledge undeserved acclaim for the rest of his life at the expense of Occidentalis!—Until the end of his days and perhaps even for many centuries after that.

At last he felt ready to undertake the return journey. He made his preparations quietly and stealthily, for fear that the natives might be reluctant to let him go. And, in fact, they were quick to notice from his behavior that something was in the wind and soon realized what was up. They continued to be friendliness itself, but a sort of body-guard was quietly set up around him, ostensibly for his own protection, but actually to keep him from carrying out his escape plans.

When he realized that any unannounced departure would be out of the question, he announced his plan openly to the head of the tribe. He said that he was extremely grateful for all the hospitality he had enjoyed and the fine and loving care he had received during his illness, but he was well again and would now have to take his leave.

The mayor of the village, as we may call him, told him point-blank that this was impossible—he had been sent down by the heavens to heal the sick in the village and there was no indication that Providence had changed its mind. As head of the tribe he would therefore actively and forcibly oppose any attempted departure of the great medicine man.

Arguments and protestations were of no avail, he was simply not allowed to go and he regretted not having been even more careful about his preparations. He was in the strictest custody, his walks were limited to "once around the block," and life became duller than ever.

As he was dejectedly reviewing his chances, a small

glimmer of hope flared up inside of him. Wasn't it possible, he asked himself, that a rescue expedition had been sent out to find him? The more he thought about it the more likely it seemed, and he decided to use this as an argument for his release. He addressed the mayor peremptorily and threatened him with retaliatory measures by higher orders if his departure was opposed any longer. Another divinity would appear, he said, more powerful than he to set him free! But all his words were of no avail.

All the same, rescue did come—not from the west, but from the east. His rival, Occidentalis, who had traveled off in the opposite direction, met him here in this distant village of savages! The moment the rumor reached the settlement that another Square was approaching, the population got ready to celebrate. They set out to welcome the new divinity with music and gifts to put him in a good mood. He was allowed to go to his colleague and the two greeted each other most warmly. A few days later they left. No one opposed their departure.

All the way back the two travelers tried to explain the miracle of how someone traveling due west could return from the east. They could not find a solution and decided to present the mystery to the club members once they were back home. Agreeing that the day of their arrival there was not the most appropriate time for such involved problems, they entered the club building with a simple "Gentlemen, here we are!" Orientalis stated simply and concisely that his friend was the winner, after which neither one said another word about the matter.

6   *The Earth Is Round*

At first the mystery gave rise to many fruitless debates within the Club of Squares, but it became a popular topic everywhere the moment the public learned of it. Everyone tried to explain the miracle, but it was only after the scientific world had become interested that an answer was found to the puzzle. It was a professor at the university of my home town, Dr. X. Pert, a brilliant scientist, who first fitted the pieces together—and his solution can be summed up very simply with a few words: "Our earth is round."

The direction upward, or "north" as we call it, the direction *going counter to the gravitational force*, is not the same for all mortals. When two plummets are suspended at a certain distance from each other, they seem to run parallel to each other, but the two directions will actually intersect at a point which, though far off, is still within a finite distance: the center of the earth.

Our earth is round! In the south there is a rocky core around which the earth's ocean stretches out, full of strange and marvelous fish. This is followed by the atmosphere, in which we first have the zone of the tropical forests. More to the outside we get the habitable zone in which the various cities and villages are located, inhabited by the civilized and also by less civilized people.

A Spaceland resident might compare us to birds, because we can move ourselves in all—that is to say, in two—directions within the atmosphere. And he might well have a point,

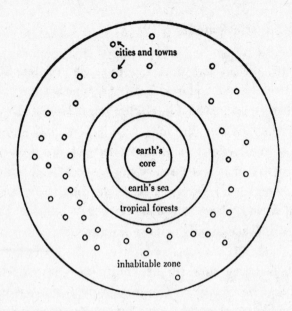

Our world disk. The outer circle represents the atmosphere's limit, which, in reality, is not so sharply defined.

even though we do not need wings to move in our fairly weak field of gravity. How this movement takes place in the horizontal west-east and vertical north-south directions will not be discussed here. Anyone interested will have to consult one of our scholarly works in biology.

Another learned professor, Dr. Newtone, discovered that gravity must be attributed to the earth's center. Its enormous mass attracts all bodies, living as well as dead. It causes raindrops to fall downward toward the core and there to form the ocean. Evaporation restores moisture to the air. A number of meteorological conditions then combine to form clouds, from those clouds rain appears—and that is the water cycle.

Of course the question immediately arises why everything is not falling down. Solid objects such as houses and buildings, and plants such as single trees and the trees in the forest, all stay put and do not show any inclination to sink. The answer is not so easy, and it might be best to just write it off to natural laws. This does not alter the fact, however, that scientific theories have been worked out to explain the phenomenon. I will be glad to touch on the matter in a few words, but this particular theory is so complicated that you need not worry if you do not understand it. Consider for a moment that all these solid objects are resting on a space parallel to our world—in other words, they are attached to a flat plane, directly beside the plane of our space. I admit that this hypothesis—it is no more than a mere supposition—is extremely difficult for a layman to grasp, even though it is not as difficult for a three-dimensional creature as it is for us. Let us therefore simply note as fact that trees and houses *do* stay put, there being no question that they do.

The scientists were soon at ease with the theory of the round world, but the general public continued to have trouble with it for a long time. They could not understand that anyone traveling due west, always west, will finally return from the east. They asked themselves whether it had really happened that way and if the Squares were not just making fun of everybody. Besides, if it did happen, why should it take such a complicated theory to explain it? The scientists had not been sitting still, however, and made clear demonstrations showing that two north-south lines drawn at different places on our world do not run perfectly parallel to each other. But as far as the general public was concerned, repetition of the trip around the world was even more convincing. In fact, such journeys are no longer unusual and are made regularly both east and west.

One problem remained unsolved, however. Does the atmosphere have an upper limit or does it continue to grow more rare, on out to infinity? No one could come up with a satisfactory answer. The arguments of the scholars and scientists contradicted each other and failed to satisfy anyone. To travel north involved insurmountable obstacles. The atmosphere becomes too thin to breathe. Serious efforts were made by specially trained sportsmen but all they could report was that the atmosphere becomes more rare as one goes higher and that visibility increases.

The trip of Aerosalta was a historic event. She was an athletic young lady who, interested in creating something of a sensation, decided to serve science in the hope of becoming famous. She certainly succeeded in the latter, but science, too, was well served by her efforts.

Aerosalta had read that it was impossible to move in the upper parts of the atmosphere for lack of air. You must understand this clearly: our organism is capable of remaining alive in these rarefied atmospheric layers, and we can even last for a short time in a total vacuum, but our movement functions are switched off, so to speak. Our girl acrobat conceived the idea of letting herself be catapulted upward by means of a specially constructed mechanism.

She began by training herself at a moderate altitude, first only at low speeds and later at greater ones. This was not entirely without danger. Even though her practice sessions took place outside of the city, innocent travelers did on one occasion come close to becoming victims of her venture. Contact with a "flying female" is of course instantly fatal. Fortunately, it turned out all right. With a swooshing speed the lady flew right between two people, who got off lightly with just a fright. By coincidence they happened to be a delegation of notables on their way to a neighboring country to settle some differences there. The first thing they thought of was an attack on their lives. An investigation was launched at once and everyone was much relieved to find that there had been no evil plot. But it was still thought best to put an end to the danger by a forceful example and the death penalty was demanded for the reckless girl.

Fortunately for her, the court was more generously disposed and gave her a life sentence instead. Her lawyer asked for clemency and made a lengthy plea. During his speech he kept playing quite casually with the catapulting instrument which was there as corpus delicti. Since the judges kept looking at it rather anxiously, he directed the mechanism, which he had meanwhile stretched for action, toward the window. With one leap his client jumped up on it, her lawyer made a startled movement which released the spring, and like an arrow from its bow Aerosalta flew out the window and into space.

Understandably, great consternation broke loose in the courtroom. Police and constables were immediately dispatched to find the escaped young woman and bring her back, but it was all in vain. Although the lawyer claimed not to have acted with premeditation and had to be believed—his profession being a foundation stone of our jurisprudence—the direction of the shot had been calculated very accurately. The space traveler had flown between the houses and the trees, landing far away in a forest where "by a strange coincidence" a resilient net had been set up between the trees.

The judges were of course unaware of this last fact, and since the lawyer had apparently been as surprised as they, not the slightest suspicion fell on him. The space traveler was now sentenced to death in absentia. The mechanism was to be seized and destroyed. Whether this happened I do not know, but it is a fact that shortly afterward the launching exercises with the catapult and safety net were resumed in a lonely area. When Aerosalta felt herself sufficiently trained she left with her personnel, including her lawyer whom she had meanwhile married, for higher regions—as far north as possible, right at the borderline where a living creature can still breathe fairly easily. One fine day, when the atmospheric conditions were favorable, she made the leap. The catapult was aimed straight north and stretched as far as possible. Aerosalta took her

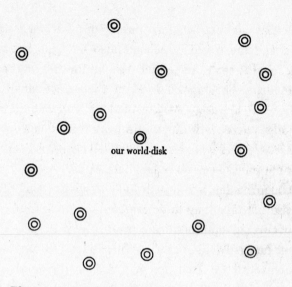

our world-disk

*There are many worlds in our space. Our own
world disk is but one of them.*

seat, and when she gave the sign, the spring jumped back
and she flew straight up toward the unknown.

According to her description, it began to grow lighter
all around her. The mist became thinner so that she began
to see objects very clearly at a great distance. Below her
the inhabited world stretched out surprisingly far away. It
was a magnificent view! But a very strange picture unfolded
above her—never seen until then by any living creature!
The firmament was not empty; at great distances and in many
different directions other worlds could be seen, some near,
some farther off, others very far away!

On her return to the inhabited world she was given
exuberant homage. The incensed judges did try to have

justice take its course, but public opinion was strongly opposed and the world of science also let it be known that nothing could come of either the death penalty or life imprisonment. The highest level of the priest ranks issued a verdict which was to be accepted as law, and it was up to the judges to see how they could work it in such a way as to save face. How they finally did this, I have long since forgotten; the public was no longer interested.

The trip by the world-renowned Aerosalta was repeated and it did not take long to determine that many more worlds exist in our universe besides our own world disk. Whether the other heavenly disks are inhabited is not known. We can philosophize all we want, but it doesn't bring us any closer to the answer. And we cannot go there because we do not have the means. Perhaps it will be possible later when science has discovered ways for transporting us through the space vacuum.

7   New Year's Eve

It was New Year's Eve and we were all gathered in a family circle while I told my grandchildren the fairy tale of Snow White and the Seven Dwarfs. The older ones, too, were listening attentively to the exciting story. I dwelt at some length on the lovely, delicate line of Snow White and I told how the stepmother had also been a striking, slim line when young and was even now considered trim of line for her age. But Snow White was the winner both in absolute size and in proportion to her length.

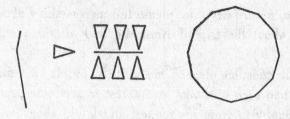

Snow White and the seven dwarfs. To the left, the witch; to the right, the prince.

My little granddaughter asked how that stepmother could turn herself into a witch so easily. I explained that she knew the art of curving her body sharply, thereby changing from a beautiful slim woman into an ugly monstrosity—and that was clear enough to the children!

My youngest grandson wanted to know how many sides the prince had. It was a difficult question. In the old stories, princes and kings only have a few sides as a rule, sometimes six or eight. But one can't come up with a six-sided prince for children who may have eight sides themselves! And yet he can't be given too many sides either. If one speaks of a forty-eighter, it will in a child's eyes approximate a priest pretty closely. I have therefore accustomed myself to giving all princes twelve sides—which is what I did this time, without any opposition.

When the story was ended, a long silence followed. But suddenly my eldest grandson spoke up: "Grandfather, tell us something else."

"What kind of fairy tale would you like me to tell you?" I asked.

"No, not a fairy tale, please tell us something about our world, about the trip of Aerosalta and all the things she saw."

That question pleased me quite a bit. It was the first time I had seen him take an interest in scientific problems. I was happy to grant his request and I told about the risky venture undertaken by our female pioneer and the marvelous world she had seen in the heavens. I explained how it had been possible after several successive attempts to remain longer at the limits of our atmosphere and to determine the direction of the other worlds.

I had an attentive audience and when I thought I had given them a complete outline of the universe with its various worlds, a general silence fell.

One of the children wanted to know whether all those worlds, which seemed to be celestial disks like ours, were inhabited. I discoursed at great length on the subject and explained that we could not know. Our telescopes were not sufficiently powerful to observe living creatures on other celestial disks, which would furthermore be enclosed by a dense atmosphere. Whether the inhabitants would be thinking creatures such as we are, triangles and regular polygons, that was very much open to question.

Suddenly my daughter-in-law remarked: "Maybe the women there are circle-shaped and the men little lines cracked in three places."

That raised a storm of protest. To suppose that women would have the form reserved in our world for the noblest form of men, and to think that men could have the shape of criminal females, hags bent in three places—that was too much for the boys.

But my daughter-in-law had a point. I explained how it might well be that other world disks were subject to entirely different natural laws dictating the shapes of their inhabitants. Perhaps the noblest creatures there were irregular polygons!

This aroused much indignation too. My youthful audience could imagine that horrible dragons or dwarfs or giants lived on other world disks and even that there might be worlds populated only by circles, but that an irregular could be a civilized, respectable creature was out of the question.

The clock struck ten—late enough for the children. After a New Year's wish, the whole troop was sent off to bed. I stayed behind with my oldest son.

After we had been sitting quietly for a while, staring into the fire, he said suddenly: "Dad, this third dimension about which your grandfather wrote his famous book, does it really exist?"

"How do you mean?" I asked.

"Well," he said, "I mean this: does that three-dimensional space with all its strange creatures like Spheres and Cubes really exist or is it fiction?"

"But you read for yourself in the book by my grandfather that he had received a visit from a Sphere, a creature out of Spaceland? What other proof do you need?"

"Yes, sure, he wrote about that visit from the Sphere, but wasn't that just something to liven up his story? Wasn't that just made up? We can visualize a third direction perpendicular to our two known directions, but does the third dimension really exist or is it only an image, very interesting and important to the philosophers perhaps, but without any reality?"

I could not say much against this. I myself had always considered the entire story of the Sphere's visit as genuine, had never doubted it in fact, but then—it was certainly possible that my grandfather had invented the story to make his argument a little more colorful. And yet I could not make myself believe it, because I could not convince myself that he would have had the courage and the strength to spend the rest of his life in prison, finally dying there, for the sake of a fictitious assumption, stubbornly clinging to an unreal mathematical concept.

"Look," my son continued, "we can imagine Lineland where all creatures are longer or shorter straight lines, moving along the straight line which is their world and on which they cannot pass each other. Such a Linelander can visualize only one direction, and when you speak to him of a second direction perpendicular to his own world, he doesn't know what you're talking about. If he were a mathematical genius he could come far enough to understand the following reasoning: When one moves a point a little, a line is established with two endpoints. If the line is now moved an equal distance perpendicular to its direction, a square results which has four angles and is bordered by four sides."

"I don't believe," I interrupted, "that a Linelander would be able to grasp that. Such a creature lacks the understanding of the concept *perpendicular* which is basic to this reasoning."

"Granted, this is only an assumption. In other words, such a mathematical genius cannot exist in Lineland. That much mathematical understanding can't really be expected from such primitive beings. But we two-dimensional creatures are familiar with the concept and the brightest among us would therefore be able to follow the argument. And so we go even further and visualize the movement of the square in a third direction, a direction we cannot observe. We can therefore conceive of such a movement, even though we cannot visualize it. The square has now been moved an equal distance perpendicular to our space. The result is a three-dimensional body, an over-square or cube, a regularly shaped object which has eight angles and twelve border lines."

"Why twelve?" I asked. This question was aimed specifically at finding out whether my son was simply mouthing the information as he had learned it or whether he could really understand it and discuss it.

"Well," he said, "the square in the first position had four boundary lines, and in the last position it also has four. Together that makes eight. In addition, the four vertices when moved described four more lines, so that the three-dimensional body has twelve boundary lines."

"Twelve *edges*, as they say in Spaceland," I interposed.

"Strangest of all, though," my son continued, "is the fact that this cube is bordered by planes, squares, six in number, of which every point, even those inside the square, is located on the outside of the solid. We Flatlanders cannot imagine that a point inside a square can lie on the outside of a body—but that's how it is. A cube therefore has eight vertices, twelve edges, and six side planes of which all points, including the inside ones, are located on the outside of the cube."

"Nothing wrong with that line of thinking," I resumed, "but can you go on and tell me what sort of object is created if the cube is now moved in a fourth direction, perpendicular to the three other directions?"

"It would become a four-dimensional body which we could call an over-cube," my son said. "Of course we cannot possibly visualize this."

"No, but three-dimensional creatures cannot visualize it either, they can conceive of it only through their reasoning, just as we can only conceive a cube but not visualize it. But let us ask ourselves for a moment by what elements this over-cube is bordered."

"Naturally it has sixteen vertices, because the cube had eight in the first and also in the new position, that is sixteen altogether."

"Right, and how many edges?"

"Thirty-two."

"Why?"

"Well, the original cube has twelve edges, the moved one does too. That already makes twenty-four. In addition, the eight vertices have each described one line, also edges of the over-cube. That makes a total of thirty-two."

"And how many lateral faces?"

"Twenty-four, namely, six from the original cube, six from the displaced one, while the twelve edges have each formed an additional lateral face during the move. A total therefore of twenty-four."

"Does that finish it?"

"Not likely, the most important point is still to come! The over-cube is bordered by eight side cubes. The six lateral planes of the cube have each formed one cube. Together with the original and the moved cubes, that makes eight cubes. An over-cube is therefore bordered by eight side cubes of which all points, also the inside ones, are located at the outside of the over-body. Of course we cannot possibly visualize that."

"Neither can the three-dimensional creatures."

"We can go on like this," my son said, "and make the over-cube move in a fifth direction, but it is no longer very clear what sort of body would then appear."

"It's still possible," I answered, "a five-dimensional, very regularly shaped body, bordered by thirty-two vertices, eighty edges, eight lateral faces, forty side cubes and ten side over-cubes. But I admit that it now becomes more and more complicated."

After that we were both silent, lost in our own thoughts. I was proud of my son, who really did have a solid insight into higher-dimensional geometry, a subject in which many members of our family had specialized. It is something of a family tradition in honor of our great ancestor, the Square, who became so famous after his death.

My son was the first to break the silence, asking: "We still don't know whether all this is simply a disconnected hypothesis, a mathematical game, or whether a third dimension does really exist. Was my great-grandfather really visited by the Sphere or is it just a well-worked-out anecdote?"

We had no idea that we would have the answer to this question so quickly.

8 The Sphere Reformed

The clock struck midnight and the great moment awaited each year by so many had arrived. You may find it strange that my family was no longer gathered in a circle around the open fire, but we thought it better for the children not to go to bed so late. We had therefore moved the ceremony up a little and celebrated the new year's arrival a few hours ahead of time. As is our custom, we had eaten the dough circles, a delicacy greatly loved by the children, but on which the grownups can gorge themselves too! I had told a fairy tale, and all in all, we thought the change of years had thus been celebrated properly. And that ended the party for the children.

The adults could come together again, but we had decided not to do this. The very instant of old-into-new is only an arbitrary one, after all, and it all depends on our own division of time and our calendar count whether this moment —otherwise no different from any other—is considered important.

So the family circle had already dispersed and I was talking with my oldest son about the third dimension. All the others had already retired to their various sleeping facilities—when suddenly it happened. I thought I heard a strange noise, a rustling or humming. Unable to figure out where it was coming from I looked searchingly all around.

"Did you hear it too?" my son asked.

"Yes, but what is it?" I wondered.

"I wouldn't know. I can't imagine where that noise came from."

"Neither can I."

This was solved at once, however. Right between us a dot appeared which expanded rapidly, becoming a circle that grew bigger. But after a while it started to diminish again, first slowly, then more rapidly, until it became a dot once more and finally disappeared.

We looked at each other, surprised but not really frightened.

"That was the Sphere," my son said.

"I think so too," I agreed.

"Then it's true!" he continued.

"Yes," I said.

So it had not been a mere hypothesis, but the truth. My grandfather had met an actual Sphere!

And now we heard a mysterious voice saying: "I did come back. I am the Sphere and I have lowered myself through your plane. Just wait one moment and you will see."

And with that the little dot reappeared, expanding as a circle, the largest cross-section of the Sphere in our two dimensional world.

For a moment both of us were completely silent. Then I remembered that I had better bid my high-ranking guest welcome, and I said: "Welcome, Your Excellency, to our world of two dimensions."

"Good day and a Happy New Year!" the Sphere said.

"And the very same to you," I replied.

"Aren't you the grandson of the Square whom I met many years ago?" my guest asked.

"Yes, that's right," I said enthusiastically. "I admire your intelligence to be able to find me among the thousands and thousands of Hexagons in our world."

"It was not easy, and I have had to eavesdrop on a great many conversations of late. In the course of this it became clear to me that there is an entirely different attitude toward the third dimension in your world today than there was before. It is now no longer a secret—on the contrary, everyone loves to talk about it even if he does not understand it in the least."

"That is an accurate observation," I said. "You could now give an open demonstration, as you did on your earlier visit, but with this difference—you would receive a hero's welcome."

"That may well be," the Sphere said, "but I don't want to do it now. And what's more, the time has not arrived yet, because a thousand years have not yet elapsed."

"That is so," I said, "but may I ask what is the reason for this visit of yours? A visit that we appreciate very much," I added hastily.

"The question is very simple. I needed it. Last time I paid you an official visit by order of the higher authorities in order to make you Flatlanders aware of the limitations of your world."

"And this time?" I wondered.

"This time I came to get something right which I did

wrong before. I have grown a little older. Though we Spheres in Spaceland live considerably longer than the Flatlanders, I shall certainly not be here when the next millennium rolls around. I am a little older and there is something on my mind which is the reason for this visit."

"Why don't you go ahead and tell us what's on your mind," I reassured him.

"If you would like to speak with my father alone," my son interjected, "I'll just leave."

"Please don't," came the quiet answer. "I have no secrets from you. You can remain here and listen too. From your conversations I concluded that you were descendants of the Square with whom I had my last discussion here and to whom I showed the wonders of Spaceland. It was easy for me to see that geometric problems interest you both, and I am convinced that you will want to hear my confession and be interested in what has happened to me."

"My son and I are all attention," I said. "We are listening."

"You will remember," the Sphere began, "as may be read in your grandfather's book—which, by the way, is also published by us in Spaceland and read by everyone interested in the properties of space—that by means of a few demonstrations I proved the existence of a third dimension to your grandfather, and after that, when he was still not entirely convinced, took him with me to Spaceland where he could look down on his own world."

"Yes, we know all about that," I said.

"Then you probably also know that your grandfather finally tried to convince me that in the series of existing worlds it is possible to go one step further and deduce the existence of a world of four dimensions. I did not want to know anything about it at that time and became annoyed. When the Square persisted, I put him back in his plane and disappeared."

A long pause followed. My son and I kept silent because we felt that we should not interrupt the Sphere in his confession. After a while he resumed his story: "The question which your grandfather posed to me was not unfounded from his point of view, and also very understandable. For him, who had just become acquainted with the existence of a dimension entirely unknown till that moment, it would be obvious to suppose that one could go a step further and assume the existence of a fourth dimension, but to me it sounded ridiculous. That too is quite understandable; I had never experienced anything unusual in that respect. Your grandfather did ask whether my countrymen had never witnessed a descent of creatures of a higher order into closed rooms. I had to admit that history books did mention such incidents but that these phenomena were regarded as visions originating in the brain of the viewer, a result of confused angularity. Now I see it differently."

Again he was silent and I could not resist asking whether he had received a visit from a four-dimensional creature.

"An Over-Sphere did visit me," he replied, "a creature from the region of four dimensions which is as strange to us as a Sphere is to you."

"I beg you," I said, "please do tell us in detail how that

came about. No doubt my son is also very interested in it."

And now the Sphere told the most amazing story that can be told to a three-dimensional creature. For us, who are already accustomed to the visit of a Sphere to our space, it was less strange.

He told how he had been sitting alone in his room when he heard a strange sound and suddenly, next to him, while the door and windows remained closed, a dot appeared which grew and grew. First it became a little sphere and finally a big one which decreased in size again, became a dot, and vanished.

What had happened? An Over-Sphere had crossed his three-dimensional world. The cross-section of such a four-dimensional body in his world is a sphere, just as the cross-section of a three-dimensional sphere in a flat plane is a circle. As we can only see the cross-section of the Sphere, in other words a circle, he could only observe the cross-section of the Over-Sphere with his space, a sphere.

The Over-Sphere performed some other tricks too. He punched the Sphere very gently inside his stomach, because Spaceland lies entirely open for Over-spaceland just as Flatland is completely exposed to a Spacelander.

The strange visitor also removed objects out of a tightly shut closet.

Then the Sphere spoke about another trick which is impossible for us Flatlanders to understand. In order to secure and fasten objects, the Spacelanders make use of ropes and chains. The ropes are tied in knots. It is impossible for us here in Flatland to visualize this, but we will be glad to assume on the Sphere's authority that it is not possible to loosen a knot in a rope if someone is holding both ends of that rope. Yet that is exactly what the Over-Sphere managed to do in less time than it takes to tell.

The other object I mentioned, a three-dimensional chain, consists of links—these are circle-shaped or elliptic pieces which lock into each other and cannot be loosened without opening up a link. But the Over-Sphere unhooked the closed links with ease.

We had been listening attentively to the interesting stories of our visitor. There were two reasons why this visit was so important to us. In the first place, it taught us that my grandfather's account of his peculiar guest was the truth, not fiction, and secondly, we learned that a fourth dimension exists in addition to the third. But it pleased us most of all that the Sphere, whom we had come to like very much, had reformed.

In a somber voice he told us how no one in his country believed him. True, he was not persecuted or imprisoned, but he was simply ridiculed and people said that he must have dreamed the entire visit of the Over-Sphere.

Our guest was obviously depressed about this and remembering the tone of our ancestor the Square, telling of his own experiences in his book, we could certainly sympathize.

"I am very happy," I interjected, "that you find a trusted atmosphere in my home where there is sympathy for you—an atmosphere which you cannot find in your own world."

"Thank you for those kind words," the Sphere replied.

"I appreciate your sympathy, but I hope that my world will soon have a better insight into these problems."

"We do too," I told him, "but if it happens we hope you won't forget us. We value your visits highly."

"Friends made in need are friends indeed," our guest said, "and they cannot be forgotten so easily."

He promised to return next New Year's Eve when we would again be able to exchange thoughts on space problems. He had a real need for friendship and was unable to find it in his own world.

We assured him that he need not wait until next New Year's Eve, but he preferred that time. He would try to convince his fellow Spacelanders of the existence of a fourth dimension and he would need much time for that. Rather than dropping in at random, he would arrive at an appointed time. Something new, worth reporting, might have happened by then.

After a warm leave-taking he disappeared from our space. The circle grew smaller and smaller and finally vanished.

We continued to talk about it for quite a while after he had left.

"It is really strange," my son observed, "that some of our insights and attitudes are more advanced than those of the three-dimensional creatures. My great-grandfather, the Square, had already achieved a proper notion about the existence of a third dimension when the Sphere was not even ready to believe in the existence of a fourth—and now that the belief in a world of three dimensions has become commonplace among us, the public in the three-dimensional world still refuses to believe in the possibility of a fourth dimension."

"True enough," I admitted, "but we mustn't forget that chance helped us here considerably. My grandfather was visited by a Sphere a long time ago, while the Sphere himself received his visit from the Over-Sphere only much later."

My son had to concede this. His thoughts wandered off into another direction, for after a while he asked: "Do you suppose there is a fifth dimension? Would the Over-Sphere ever be visited by such a creature, much more highly developed, from the world of five dimensions?"

"That could well happen," I said enthusiastically. "Why should a mathematical process—for isn't that what this really amounts to—ever cease? In fact, there might well be an infinite number of dimensions!"

"But the fact that we can conceive of multidimensional spaces is no reason for them to exist," my son pointed out.

"That's a good point," I agreed, "because only after receiving signs of life from a world outside of one's own can one conclude that such a world exists. Until then, assumption of any unknown world is nothing more than a fantasy."

"Yes, but a lovely one," my son declared. "Perhaps it would be far from pleasant for us to meet creatures living in worlds of many more dimensions. They would think us very primitive and look down on us."

And so we talked far into the night. It was only a daydream and it did not bring us any further, but it had done me good to find my son to be someone with whom I could exchange ideas on these problems so easily.

PART II

Congruence and Symmetry

9 Pedigrees and Mongrels

During the next year I often found my thoughts wandering to the Sphere's visit. How was he? Had he talked in his world about his visit to Flatland? Had he continued his efforts to give his fellow beings a little more geometric insight? Had he found an audience? Or might he perhaps have been thrown into jail for his revolutionary ideas? Then he would not be able to come and see us at the appointed time.

One evening, after the others had already retired, I was still discussing a number of subjects with my oldest son, when the conversation almost automatically turned to the topic which interested us both so greatly.

"I don't believe," my son said, "that the Sphere is a prisoner, because for many centuries now efforts have been under way in Sphereland to convince us Flatlanders of the existence of several dimensions."

"That is true," I answered, "so far as they are anxious to convince us of the existence of a third dimension, but not where a fourth dimension is concerned, in which they do not believe themselves. If our friend the Sphere has been careless enough to tell his fellow spacemen about the existence of a fourth dimension, it is very possible that he has been put under lock and key, and we won't be able to rescue him."

"You are right," my son continued. "The Over-Sphere is the only one who could set him free without any interference on the part of the Spacelanders, because prisons are open toward the fourth dimension."

"Let us hope that nothing serious has happened, and that the Sphere will be looking us up on New Year's Eve. Even though there may not be anything new to talk about, we can still renew our acquaintance and have a pleasant conversation about the geometry of many dimensions."

"Yes, not much important is happening here," my son said, "at least, not much that would interest the Sphere. I don't think we can expect him to be very interested in Agatha and her pups."

To make this clear, I will first have to explain what "pups" or "dogs" are, because you have never heard about such animals from me, nor was any mention made of them in my grandfather's book.

There were no animals anyway, at that time. At least, no domestic animals. I have told you that during the time of the exploratory travels, the southern jungles were discovered and also the world ocean which is located in their interior. The jungles were populated by strange creatures, all of which we have loosely called "dragons," though biologists have distinguished and named many different species among them. Even stranger creatures live in the world ocean and not all of them have yet been studied.

Now it is not right that animals should come to be known only as a result of exploratory trips, because in ancient stories, legends, and fairy tales, dangerous animals occur

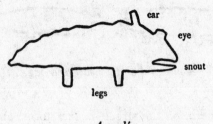

A wolf.

which are called either dragons or wolves, and it was well known that they lived not only in the southern regions, but also in the woods and forests of more temperate zones, to be entered only at one's own risk.

The common wolf is an animal with one snout and two appendages called legs, though their real function has not been determined; they are not needed for movement. It furthermore has two bumps: the one nearest to the snout is the eye and the other one is the ear. A wolf is a dangerous animal; it can wolf us down in one bite.

Years ago, enterprising men captured young wolves to exhibit at circuses and fairs. When the animals were fully grown they were dangerous. Then they would be locked up in strong cages so that the public could view the beasts of prey without danger.

As a result of selecting the most tame animals and breeding them, a species gradually evolved which was so docile that it could be used as a domestic animal. It was called a "dog."

Dogs are tame and tractable, and can be easily kept in the living room. They sleep in a corner of the passageway or even in the bedroom. They follow their "master" or "mis-

A mongrel and a pedigreed dog.

tress" everywhere, on the street, in the stores, and on walks.

But to get back to Agatha, she is my only daughter, a nice girl, slim and pretty. She had a dog, a girl-dog, and she was just crazy about it. It was not really an expensive possession because it was not pedigreed, but a genuine mongrel. No difference in form or shape is involved here, a pedigreed dog and a mongrel can be equally graceful and elegant, it is a difference in "turning sense."

I will make this clear to you. When I walk around a mongrel in the positive direction, that is to say: first to the north, then to the east, then to the south and finally to the west, I successively pass the ear of the animal, its eye, its snout, and its legs. If, on the other hand, I walk around a pedigreed pup in the positive sense, then the succession is: ear, legs, snout, eye, ear. A pedigreed dog and a mongrel are mirror images of each other. They are "symmetric," but not congruent, as the experts call it.

No matter how one turns a mongrel, one can never turn it into a pedigreed dog; it remains a mongrel. From the day of his birth a dog is either a mongrel or a pedigreed dog and so it remains.

As a result of an accident of nature, only a few pedigreed dogs exist in the world. There is barely one for every thousand street dogs. Because of this, a pedigreed pup is

No matter how a mongrel is turned, he cannot be turned into a pedigreed dog!

thought much more handsome and noble, and sometimes exorbitantly high prices have to be paid for it. Now you can say that this makes absolutely no sense, but fashion does not let itself be influenced by reason. A pedigreed pup is rare and therefore expensive. A lady who is walking a pedigreed pup is a chic lady. Everyone watches her.

Agatha's pup was a mongrel. She loved the animal and was very happy with it, and would probably have stayed so if another girl had not lived in the same street who was called Chromosa and who had a pedigreed dog. And even that could have been borne had it not been for the fact that a nice young man, a Pentagon, who had been paying her a great deal of attention, now suddenly turned his attention away from her and concentrated it on Chromosa, whose dog had become the mother of seven puppies, three of which

Chromosa out walking with her dogs.

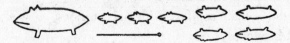

were genuine pedigreed pups. How proudly Chromosa walked down the street with her noble pedigreed dog train!

Agatha was very sad. I tried to comfort her and said that the boy was not worth it if he looked more at the dogs than at the girl, but this did not help. I therefore decided to do my daughter a special favor.

I went to a good acquaintance of mine, an Equilateral, who had a male pedigreed dog, and after some negotiating I got what I wanted. It cost me a little more than I had anticipated, and I must honestly say that my friend did overcharge me considerably, but I had made up my mind to do my child a favor. The upshot was that shortly thereafter we had a visit from my acquaintance and his dog.

We waited out the necessary weeks in great suspense. A pedigreed pup is a rarity. Among the puppies of a father and a mother who are both pedigreed dogs, at most 30 percent are pedigreed, and in this case the mother was a mongrel!

Finally the time had come. Agatha's dog brought twelve puppies into the world, but . . . there wasn't a single pedigreed puppy among them.

It goes without saying that Agatha was terribly sad, and I was no less so. I really sympathized with her. The young man no longer looked at her at all and a short time later he married Chromosa. The pedigreed pups followed her all the way to City Hall and one can imagine that the procession attracted a great deal of attention.

Agatha sought comfort with her puppies, to whom she gave all her love.

10 *Little Red Riding Shoe*

Once again it was New Year's Eve. The fun for the children started early and they feasted on the fried dough circles baked by my wife and daughter-in-law. Naturally, Grandfather had to tell a story. I had chosen the fairy tale of Little Red Riding Shoe for this purpose. They had all heard it many times before, but it is always nice to hear it told once again, and the fact that it was New Year's Eve made it all the more exciting.

"Little Red Riding Shoe," I began, "was a sweet little girl, but not very obedient. On her birthday her parents gave her a little red shoe which looked awfully pretty and which she wore all the time. All the people of that little village therefore called her Little Red Riding Shoe."

"Was she really allowed to wear that little shoe *all* the time?" one of the youngest ones, who had been listening very closely, wanted to know.

"Certainly. Her mother had told her it would be perfectly all right. Why not?" I said.

"But then why was she disobedient?" the little one wondered.

"Oh, you'll find out in a minute," I reassured her. "Just listen.

"One day her mother told her little girl: 'Little Red Riding Shoe, listen here. You know that grandmother has been sick for a long time. She is a little better now and I have fixed her a basket with these good dough circles and a little bottle of wine. Now I want you to go over and take it to her.

Ask her how she is and wish her a quick and complete recovery.'

" 'All right, Mother,' Little Red Riding Shoe said, 'I'd love to.'

" 'But listen carefully,' the mother went on, 'be sure to take the main road which everybody uses. It is a little longer, but the forest is too dangerous, not because you may get lost, but a wolf lives there and that is a very dangerous animal.'

" 'All right, Mother,' Little Red Riding Shoe said, and off she went. It was a lovely day and she was happy to be able to take this long walk all by herself. Usually, when she went to visit her grandmother, her father or mother would go along, but they did not have time today. Mother had to stay at home because she had to get the food ready, and Father was away at work. He was a woodcutter and worked in the forest where he cut the trees from which they made shelves. It's true that he was not really fitted for this job because his top angle was 10°—a little too much for a woodcutter—but his father and grandfather had also been woodcutters and that is how he had come to have the same job. If he ever had a son with an even less sharp top angle, he would have to find him some other type of work. He thought about it often but there was no point in worrying about it ahead of time. Right now he had only one child, a little girl, Little Red Riding Shoe, a nice child, and he and his wife loved her dearly. But she was not always obedient!

"Little Red Riding Shoe walked on very quickly until she reached the fork in the road. The main road circled the forest, it really was a longer way around; a smaller path went straight through the woods, right up to the little house where her grandmother lived.

"Mother had said that she was not to go through that forest, but was that really necessary? Why should she have to make a detour? She had gone right through the forest so often with her father and nothing had ever happened. She had never seen the wolf. Would he really be there today? Perhaps he always took his afternoon nap at this time, and today especially, because it was pretty hot. And wouldn't he be afraid of her? She had an awfully sharp point, after all. The wolf was of course very dangerous for little boys of nobility who could not defend themselves so well.

"While she was working things out in her mind this way to explain away her disobedience, she had already traveled quite a way along the narrow path leading straight up to the forest. After reaching the first trees she hesitated a moment but to turn around . . . after all this time . . . no, now she'd go straight ahead. If she walked quickly she would be there very soon. And with this firm resolution she was inside the forest.

"She walked on, looking all around to see if she could spot any woodcutters, because she was becoming a little frightened after all. You could never tell what kind of animal might suddenly jump out from behind trees. But she did want to be brave. Although she would not tell them at home, not right away at any rate, that she had gone straight through the forest, she would certainly brag about it a little to her girl friends. She wanted to be brave and resolved that, come what may, she would not show that she was a little scared. She would put up a good front if anything happened!

"And something did. She had just reached the heart of the forest when the wolf was suddenly standing right in front of her. She had not heard him coming. How terrible he looked! So irregular and rough of surface! She was terribly frightened!"

"She should have run away very, very quickly," one of the children called out.

"I would have run right back home," another said.

"That would have been the best thing to do," I continued, "but Little Red Riding Shoe was not that bright. The wolf said in the sweetest and most flattering little voice: 'Hello, little girl, where are you going?'

"When the wolf spoke to her that way, Little Red Riding Shoe was no longer frightened. She even thought the wolf was awfully nice and told him exactly where she was going: 'I am going to my grandmother who lives at the other end of the forest. She has been sick and I am bringing her some fine dough circles which Mother baked for her.'"

With great interest the children followed the story, which they already knew very well but which they continued to find exciting though they were hearing it for the umpteenth time. I told how Little Red Riding Shoe entered a contest with the wolf to see who would be first to reach the door of her grandmother's little house, how the wolf was able to trick his way into it, and how the grandmother was terribly afraid when she saw the savage beast in front of her. "Her fright did not last long though, for the wolf gulped her down in one bite."

It's strange that this did not bother the children at all. But then, they knew how it would end and could furthermore be sure that the grandmother would come out again alive. They did think that Little Red Riding Shoe was awfully stupid not to see that it was not her grandmother but such an irregular creature as a wolf who was lying in bed.

" 'But Grandmother, what happened to your slim figure?' Little Red Riding Shoe cried out.

" 'That is because of the disease, my child, which has made my body swell up so irregularly,' came the quick answer of the deceiver. When the child finally realized her mistake, it was too late. She disappeared into the stomach of the hungry monster. But in that process she had fortunately lost her shoe."

"Why fortunately?" one of the boys wanted to know.

"That's clear," a little girl said, "now she was sharp at both ends."

"Right," I said. "Little Red Riding Shoe finally understood the danger threatening her and the wolf realized too late that it is almost always fatal to gobble up such a young and lively little girl. That she was vibrating with anger was dangerous enough, but she had no sooner entered the wolf's stomach when she began a violent movement, back and forth, while pricking into the stomach lining in all directions so that the beast bellowed with pain.

"Even that would not really have helped her if she had

not been quick-witted enough to understand that she should direct her vibrations in one direction. As a result of her repeated jabbing pricks, a hole developed and before she knew it, she was outside the body of the dying animal."

"And the grandmother? Could she find the opening?" a little one asked sympathetically.

"No, Granny was no longer able to do that. Her eye was in too poor condition and the time spent inside that tight little space had dazed her almost completely. Fortunately, however, all the roaring and bellowing had been heard and Little Red Riding Shoe's father came running up with some foresters. With their top angles they split the monster into pieces and helped poor Granny out into the open air and sunshine just in time.

"Little Red Riding Shoe, who had immediately put her little red shoe on again, told at great length all that had happened. Her father was so happy to find his child alive that he forgot completely to scold her for her disobedience."

The story was ended. The girls in particular were satisfied, for isn't the fairy tale of Little Red Riding Shoe really *the* fairy tale for girls? Cinderella and Snow White are famous and important, but they do look up to a multisided prince as their lord and master. Little Red Riding Shoe, on the other hand, knows how to defend herself—and that ably and resolutely—against a horrible monster. In fact, we might well regard Little Red Riding Shoe as the patroness of the feminist movement.

But I did not let this get to the debating stage because it had already struck ten and I sent the entire outfit to bed.

11 A Magic Trick

After the smaller children had been put to bed, the older ones stayed up in a cozy group. A large circle was formed, consisting of my wife and the oldest children. Agatha had her dogs with her. There were twelve pups, actually too many to keep inside one house! My second son, who can be a little rough, sometimes said: "Everywhere you go you fall over Agatha's mongrels. It's just plain ridiculous to keep so many pups. We should do away with ten or eleven of them."

But then Agatha would make loud protests and I must admit in all honesty that I have a tender corner in my heart for my daughter and in spite of all the fuss and bother they make, I do not want to deprive her of any of her pets.

We were waiting for our visitor, the Sphere; even though it was not certain that he would be able to make it, in our hearts we were convinced that he would be there. The nearer it approached midnight, the greater the tension grew.

"He could be sick," said my oldest boy, without saying whom he had in mind, but we all knew he was talking about the Sphere.

"Perhaps he is dead," another said.

"Or in prison," was the opinion of the third. But all tension melted away when, at the last stroke of twelve, a small circle appeared. This was the beginning of the arrival of our friend the Sphere in our space.

We let the little circle quietly grow until it had reached its largest dimension, well known to us from before. I waited a moment longer and spoke: "Welcome, Your Excellency, in our midst. We welcome you as a friend."

"I know that," the Sphere said. "I understand and I feel that. In your midst I feel myself among friends. Of that you may be certain."

It was indeed very clear from the friendly tone which the conversation took. The Sphere asked about the whole family and wanted to know whether any important events had occurred in Flatland during the past year.

My wife and I took turns relating various small incidents which had taken place, more in order to maintain the cozy atmosphere than because we believed they were anything that could interest our friend. At one point I did think of telling about Agatha's puppies and her distress over the fact that all of them were mongrels, but I did not want to stir all that up and open old wounds for her.

The Sphere did not have much to tell either. After a few reports of little interest we were talked out. The visit was more like a common courtesy call than an exchange of thoughts about scientific subjects.

Nevertheless, the conversation was to take an important turn. I had noticed that our guest had been watching one of the puppies for quite some time. All of a sudden he asked whether that one was a young criminal. This comment produced great hilarity. In order not to have it seem that we were laughing at the Sphere because of his comment, I immediately started to speak and said something as follows: "It is certainly understandable that you would make such a comment because you know that in our world the advanced creatures are all regular polygons."

"Or straight lines," my wife quickly added.

"Of course," I said, "as you know, our women are straight lines. The less-developed men are isosceles triangles and all irregular figures are criminals. But these small creatures you see here are domestic animals—the descendants of wild animals tamed centuries ago, many of which still inhabit our southern forests."

The Sphere seemed to follow my explanation easily, because he answered that Spaceland too had irregular creatures, the domestic animals. Of these many different kinds were known, such as dogs, cats, sheep, pigs, horses, cows, and goats.

"Do you," my son asked, "keep all these animals in your house?"

"No," he said, "but there is a domestic animal of another sort, a queer creature which has a very irregular form."

"In other words, a criminal," my son noted.

"Perhaps," the Sphere said, "although it does not consider itself as such. It calls itself 'Homo sapiens' and believes itself to be a highly developed creature. It is a symmetric animal."

In order to give us some impression of this strange animal, the Sphere placed the blocks from my grandchildren's building-blocks set in a certain form which all of us examined very carefully.

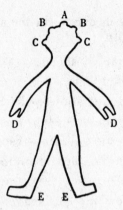

The Sphere lays out a figure which is supposed to represent Homo sapiens. (A mouth, B eyes, C ears, D grasping organs, E propulsion organs.)

"Doesn't this monster harm anyone?" my second son asked.

"Yes, it does," the Sphere answered. "It considers itself justified in killing and eating lesser-developed creatures, or putting them in prison, or exterminating them, as it thinks best."

"And it still is not a criminal?" one of the others ventured.

"It does not consider itself to be such. It has great thinking capacities," the Sphere replied.

"But it must certainly live in lasting peace with its own fellow beings," Agatha said.

"That is not true either," the Sphere said again. "These animals are constantly waging violent, bloody wars against each other. Large groups are driven to their deaths by others and sometimes tortured, tormented, and murdered with all possible means."

"Then we must agree that they are the greatest criminals

of all," one of us said, and we agreed. Even the Sphere had to admit this to a certain degree.

"Actually, I cannot deny it, but Homo sapiens has high ideals which it hopes to realize."

"And what excuse does it give for its misdeeds?" Agatha wanted to know.

"Well," the Sphere said, "they have all sorts of excuses. They always blame each other for the big wars, and when undeniably personal misdeeds are concerned, the reason given is like this: 'I had to do it to prevent something worse.' "

"I don't see," my oldest son declared, "that we have any reason not to consider Homo sapiens a criminal without a conscience. And his form indicates it too. He may be symmetric, but he is very irregular for all that!"

"And the women among these creatures, are they straight lines?" my wife wanted to know.

"No," the Sphere said, "they differ only very little from the men in their form."

"But how can one distinguish one from the other?" was the counterquestion.

"In the first place by their clothes. Except for the heads and grasping organs, these creatures cover themselves entirely with pieces of cloth," the Sphere informed us.

A silence followed. This news needed first to be digested. Agatha was the first one to break the silence, by asking: "Do the females of these creatures also wear shoes?"

"Certainly," the Sphere said, "and not only the females, but the males too. The shoes of the females are usually more graceful and elegant."

This reassured Agatha a little. One of my sons then asked: "Does such an animal wear one or two shoes?"

"Two," the Sphere assured us, "one on each foot, a left shoe for the left foot, and a right for the right foot."

This was not immediately clear to everyone.

"Look," the Sphere said, "the feet are symmetric. They are the mirror images of each other, but they are not congruent. No matter how you turn or twist a right shoe, it will never become a left shoe."

"In the way a mongrel dog could never become a pedigreed one," my son thereupon observed.

He was right, of course, but I did not think it was nice of him to say it so bluntly. He could have used another example. Agatha had a tear in her eye. The significance of this observation had escaped the Sphere, who naturally knew nothing about Flatland's mongrels and pedigreed dogs. But he did notice that our attention was suddenly riveted on the puppies.

"Look," he said, "here you have two little dogs which are completely identical. They are congruent."

"And they will remain congruent no matter how they are turned," my son added.

"That is to say," said the Sphere, "if they are not lifted out of their space, out of the flat plane, because they can certainly be turned around through the three-dimensional space."

This was too much for us to understand. The Sphere therefore resorted to action. He took hold of one of the puppies and whatever he did with it none of us could

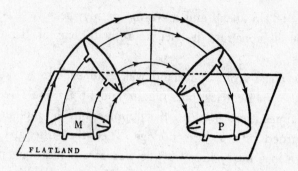

By lifting the little mongrel out of Flatland and turning it around, the Sphere turned it into a pedigreed pup.

follow. Now, afterward, it is clear to me that he lifted the little animal out of our space, turned it around with an amazing swing and brought it back to Flatland. The little mongrel puppy had become a pedigreed puppy.

"Go ahead and compare it with this other pup," the Sphere said. "They are no longer congruent, but they are symmetric."

At first Agatha was struck dumb with surprise, but suddenly she gave a shout of joy and cuddled and hugged her pet most warmly.

The Sphere who thought that she was very unhappy, told her: "Don't worry, little girl. I shall bring it back to its old condition."

But that made Agatha protest with every bit of strength she had in her. Of course the Sphere did not understand the cause for all her excitement and was not sure whether he had done something wrong or not. I hastened to reassure him and said: "Go ahead and leave the puppy the way it is, but can you demonstrate it once more with one of the other puppies?"

Our guest from Spaceland was kind enough to demonstrate the magic trick once again and at Agatha's request eleven times more until all the puppies and the mother had been turned into pedigreed dogs. Then the little girl said goodbye in a hurry because whe was afraid that the Sphere would continue his experiments. She disappeared with all her dogs, pretending that she wanted to go to bed, but actually she could not sleep a wink the whole night. What would all her enviers say when tomorrow, on New Year's Day, she would take a walk with her whole trail of pedigreed dogs!

We talked a little while longer about this question of symmetry and congruence.

"If I understand it correctly," I said, "isn't it possible for something like this to happen in Spaceland?"

"Impossible," the Sphere answered. "A right shoe remains a right shoe, no matter how it is twisted or turned."

My learned son now entered into the conversation and said: "But that can't be true. If that right shoe were to be lifted out of its three-dimensional space and turned around through the four-dimensional space and replaced in its origi-

Agatha out walking with her train of pedigreed dogs.

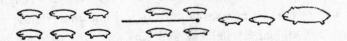

nal space, then it would be changed to its mirror image and become a left shoe."

"I cannot conceive how that would be possible," the Sphere said.

"Of course not," my son replied, "just as we cannot conceive how that swing through the space of three dimensions took place. You might ask your friend the Over-Sphere to perform such a feat."

"I don't believe it is possible," the Sphere repeated. "That can happen in Flatland, not in Spaceland."

He took his leave and vanished.

12 Vision of Lineland

I do not need to tell you how happy Agatha was. Next day she took all her dogs out walking and attracted a good deal of attention. Admiring or envious glances everywhere! I did not begrudge her this, although I do think it is very foolish to judge someone according to the directional turn of his dogs!

Otherwise life continued placidly. The only topic of discussion for us following the Sphere's New Year's Eve visit was the possibility of reversal as the result of a swing through the area of three dimensions. Of course we cannot visualize this, but we can understand it a little better if we let our thoughts take us to Lineland.

It came over me in a dream. I cannot call it coincidental

—since my mind had been occupied with the problem for days now, it is not at all surprising that my tension was expressed in the form of a dream.

And so I had a very clear dream vision of Lineland. I could see it before me: a long line with all sorts of line-shaped creatures, all lines, long and short. The king in the middle. To the left and right of him men, women, and children. Gaily they vibrated back and forth in their one-dimensional space. Of course they were unable to pass each other, bound as they were by their space, so limited and yet infinitely big at the same time. Yes, infinitely big, even though it was only one single line!

I spoke to the king, who cried out angrily: "Where does that voice come from? Is that a magician addressing me?"

"No," I called, "that is the voice of a more perfect creature from the world of two dimensions. My voice is reaching you from outside your own space!"

"I don't believe in supernatural things," the king called back. "Either I am dreaming, or I am suffering from hallucinations."

"Neither is true," I replied, "but I shall come to you, in your space."

Putting my words into action, I eased myself down into his space.

"How is it possible," the king cried out, "that a huge creature suddenly arises which was not here before? It's magic and nothing else!"

And with those words he hurled himself at me. I managed to disappear from his space in the very nick of time and with a great crash he ran into his neighbor, who might have been a victim of his king's violent emotional reactions on other occasions, but who had probably never suffered quite such a collision before.

"May I ask you something, Your Excellency?" I said in a soft but firm voice.

The word "Excellency" seemed to soothe the king. He coughed slightly and spoke: "I will grant you a few words."

I did not want to risk spoiling his good mood and phrased my sentences as politely as possible.

"Your Excellency," I said, "you have no doubt found yourself occupying the same spot in your space for many years now, with many men, women, and children to either side of you. On one side I see first two men, then a woman, then three boys, and on the other side first a man, then three women, then another man, and then five boys."

"Yes, I know that," the king interrupted me. "The smallest child can tell me that."

"I understand, Your Excellency," I continued, unperturbed, "but how about my turning you around?"

"Turn what around?" the king called out.

"Turn you in such a way," I explained, "that on the side you now have two men and one woman, you will have one man and three women, and on the other side on which there are now one man and three women, you will instead have two men and one woman."

"Impossible," he snapped back.

"Watch out," I said, "it is happening."

Whereupon I lifted the king out of his line-shaped world,

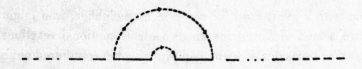

How the king of Lineland was turned around.

turned him around, and simply placed him back in his line.

"What's happening to me now? Stop it, magician!" he cried out.

"It has happened," I said. "How do you feel?"

"Terrible, just plain terrible," he said. "You jarred my insides with all your tricks. I am hurting all over!"

"Keep quiet," I admonished, "it will pass quickly." And it is a fact that the outburst I had expected failed to come. The king remained reasonable.

"Where am I actually?" he asked.

"In your own space," I assured him.

"My own space? My own space? There is only one space and that is mine; you are right about that, because I am the king. But it is all so strange." And he immediately put his two voices to work which were located at each end of his body. He was trying to establish sonic contact with his subjects. We might say he was "calling the roll." Each in turn, the residents of this line-shaped world put their voices into action.

"What has happened?" the king complained. "The world is turned around! What was on my right is now on my left, and what was on my left before has now turned up on my right. How can that be?"

This surprised me. I had turned the king around in his own world and now it seemed that he was totally unaware of this and thought it was not him but the entire world which had been turned around.

I was puzzled. What could I tell the king? First of all, I had to get the matter straight myself. And so I was standing next to the hustle and bustle and the movements back and forth of all the Linelanders. Everything clouded over as if the whole little world had evaporated into a mist. I awoke and to my surprise found myself in my own bedroom, in my own two-dimensional world.

For a long time I walked around with this problem which the king of Lineland had posed in my dream. When the right moment presented itself, I talked it over with my son and discovered that he found the reaction of the king of Lineland perfectly natural. The king, whose body came to lie in the opposite direction, now received all impressions from precisely the opposite side as before. Voices of people to the east of him, for example, now reached that ear first which formerly heard them last. Clearly it would appear to him as if these persons were suddenly on the west side. As he now saw the world, it was not him, but the world which was turned around.

Still, I must admit that I was not convinced. It is true that my respect for my son's mathematical insight was such that I wanted to believe he was right—and which I now know he is—but other things had to happen first to make me realize it.

With all this, our family circle had enough to discuss. Whenever we had company, the talk would turn to these strange events—which stands to reason, in view of our pre-occupation with them—and the starting point would usually be someone's question, how Agatha had acquired such a magnificent set of pedigreed dogs.

Agatha had wanted us to give an evasive answer to the question. She gave several reasons. In the first place, she was afraid the value of the pups would be questioned if they were found to have come into being, not by birth as is customary, but by what she called a scientific magic trick. Secondly, she feared that after a time the prescription for turning mongrels into pedigreed dogs would become common knowledge with the result that she would no longer be the only privileged one to have profited from it. There was probably a third reason which she did not mention—namely, that she thought it more impressive to act mysteriously. The world would just have to think that the pups had cost a small fortune. This pleased her vanity more than spreading the story of a mysterious magician who had paid us a visit.

However, my open nature rebelled against glossing over the truth. Yet the reaction was different from what I had expected, as evidenced by the incredulous faces of our guests. They clearly had their own opinions about the matter, that is to say, they thought we did not want to give the real reason. If we had a magician among our friends and acquaintances, a magician who showed up only in the dead of night on New

Year's Eve and whom no one else had seen, why didn't we simply ask the man to conjure up great riches for us instead of letting him perform tricks with little dogs? What was the real reason we did not want to reveal how we had suddenly become rich? This was, after all, quite suspicious! No wonder that a lot of talk was making the rounds about us and that people turned around to stare at us on the street. Whenever we entered a hall filled with people, talk would suddenly die down. We could sense that they had been talking about us. All in all, we were company it was best to avoid!

The ones most kindly disposed attributed the origin of the expensive pups to black magic. Yes, it might be that the nocturnal visitor who was behind it all had really been the devil in person, and people who associate with Beelzebub had better not be counted among one's friends!

13 The Vertato Case

A few months later I chanced to talk with an acquaintance who is a neurologist. As doctors usually do, he spoke of an interesting case that had turned up in his practice, a patient suffering from a delusion never before recorded in the annals of medical history. When this man awoke one New Year's morning, the whole world appeared to him to have been turned around. His bedroom looked very strange to him. The table was standing exactly opposite from where it usually stood and the door, too, seemed to have been moved. When he walked out in the street he noticed that the vehicles were keeping to the wrong side of the road, but what interested him most was that all inscriptions were in mirror script. He thought that he might be having a dizzy spell, or that he was dreaming, but this was not so. He felt very miserable; this strange world disturbed him. How did everything get to be so strange all of a sudden?

He tried to think it through calmly and quietly. Was he ill? Very ill? Suffering from an undiagnosed disease perhaps? He cut his walk short and returned home where he went quickly to bed. When he closed his eyes he felt normal again. Time and again he wanted to get up because he was completely rested, but then he would see the objects around him in the room, just as always and yet so completely different, and he decided he'd better stay in bed.

The next day the situation was no different but he could not stand it in bed any longer. How difficult it was to find his way in his own home! Every object was precisely on the side where he did not expect it. With great difficulty he reached his office. The side street through which he had to pass was on the wrong side, the door opened the other way. Everything was just as strange as in his bedroom.

But now it grew even worse. The letters waiting for him were illegible. They were written in mirror script!

He sat down and began to work, wrote an urgent business letter, and handed it to his secretary. She took one look at the writing and stayed right where she was, looking at her boss in stunned amazement.

"What are you waiting for?" he asked.

"I can't read this letter," the girl said.

He took the letter from her but could find nothing strange about it. The letter was perfectly legible, as usual. He was more puzzled than ever. Was he sick after all? Perhaps a serious mental disturbance? Maybe he had better go back home. If it had not gone away by tomorrow, he would call a doctor. He went home, but it did not go away, not that day or that week, not at all, in fact. He became so confused that he had to be admitted to an institution for the mentally ill.

I had been listening to this report with mounting interest. My explanation was ready: The man had been turned around as the result of a swing through the area of the third dimension.

I suggested this hypothesis, but my medical friend rejected it and from his slightly mocking smile I perceived that he did not think me quite normal either. Of course he had heard the talk about our family and it was obvious that he thought himself to be dealing here with someone who was showing a dangerous mental deviation.

He refused my request outright for permission to visit this patient, and I did not dare press it too urgently. I did ask about the sick man from time to time and would hear that things were going well with him. He had already learned to read and write again in the usual way and after some time he was discharged from the clinic as completely cured and able to resume his normal activities.

As for myself, I was convinced that this was a similar story to that of Agatha's pups, and my son, whom I had of course told the story in every detail, agreed with me. The only question remaining was how this turnabout had taken place. We tortured our brains in vain to explain it, and after intensive deliberations, we decided that my son would visit the man in question. We now knew his name; he was Mr. Vertato, a well-known businessman who lived in our own neighborhood.

My son thought it better if I did not look up Mr. Vertato myself because of all the talk about me. I agreed and we decided he would go instead.

Nevertheless his visit did not shed much light on the incident. The patient claimed that he had felt nauseated and dizzy on New Year's Eve, as if he were suddenly seriously ill. He had therefore decided to stay in bed all the next day and the day after that had returned to the office as usual where he had encountered the startling experiences described above. He told how it had been necessary for him to learn to read all over again. Word images were in reverse for him and he had to write the same way. He had reached the point where it no longer gave him any trouble. The result of all this had been that he now could read mirror script as easily as ordinary writing! But most important of all—he had regained his zest for living.

"I can understand very well," my son said, "that you found the world strange, or rather that you had the feeling of no longer being at home in the world, of no longer belonging in it."

"That's it," Vertato answered. "I felt as if I were in a strange land, and I had to get used to such strange surroundings, my own home, my own office, it was even difficult for me to find my family, my friends and acquaintances in the persons around me. It took time."

"But fortunately it turned out all right," my son remarked. "Was it a long, slow process which restored your courage to live or did it happen all of a sudden?"

"All of a sudden," Mr. Vertato resumed. "It was because of the music. I don't know where these sounds came from. Outside? Next door? But suddenly the notes of a well-known piece of music registered in my consciousness and all of a sudden I understood that there was something in this world which had not changed, the succession and combination of tones in the music came to me exactly as they had done before. I never enjoyed a simple musical composition so much as I did this time. It was the turning point of my illness. It gave me the courage to face life again."

At home we listened with much interest to my son's account of his visit. We were struck by the fact that the strange event had occurred on New Year's Eve, when we were to have had a visit from the Sphere. Was there any connection? This question was to engross my son and myself the rest of the year.

14 Experiments in Spaceland

It was for us, exiles as we were from society, a real celebration to gather in a family circle on New Year's Eve. It was out of the question that anyone would drop in on us and for this reason we could be grateful that our outer-space friend the Sphere used to visit us on that particular evening.

We longed for his arrival and the tension mounted as the clock approached midnight. Of course our beloved fried dough circles were passed around again earlier in the evening, and as paterfamilias I had to tell another fairy tale. This time I chose The Sleeping Beauty, a princess who lived in a big castle that stood in the midst of an almost impenetrable forest. The trees in this forest were unusually prickly and they grew so thick and heavy that it was impossible to walk between them. It took more than a hundred years before an enterprising prince, a twelve-sided one, of course, began to try to penetrate the forest. It was never clear to me how he managed to do this. I usually said that the trees had begun to shrink quite a bit with age, a decay phenomenon which does occur in nature.

The prince found the princess and the entire family, including the staff, fast asleep. Apparently the crowdedness of the trees had cut down the air circulation, thereby hindering the supply of fresh air needed for breathing, and a general state of sleep had resulted. The special scents of certain plants in the castle gardens had kept the occupants of the castle in a miraculous state of twilight sleep and they did not die.

The prince was immediately enchanted by the slim princess. He kissed her and thereby blew fresh air into her lungs so that she awoke. The supply of fresh air was now just sufficient to awaken all the people from their hundred-year-long sleep. A wedding followed. The forest was naturally thinned out, the castle was spruced up, and the young couple lived long and happily.

The time had come to send the children to bed. But my oldest grandchild was allowed to stay up and be present during what was still to come.

We waited for the big moment on veritable tenterhooks and fortunately were not disappointed. On the last stroke of twelve a small circle appeared which filled out until the Sphere, his largest cross-section in our space, was back with us.

Even though I of course addressed him as "Your Excellency," we considered him a friend of the family and apparently he too felt very much at home. The conversation quickly turned to reversing people by a swing in a space of more dimensions and I told him the story of the businessman for whom left and right had suddenly become reversed, which we knew as the case Vertato. He was very much interested and told us that he himself had turned the gentleman around. Right after leaving us that evening, he said, it ccurred to him that it was a pity to limit the reversal experiment to a little puppy which was unable to communicate its experiences. If he had thought of it earlier, he would have turned one of our family around, but it did not occur to him until he had already left our space. He simply carried out the experiment with the very first man he met.

He could not say exactly where the house of his victim was located. It could not have been far from our home, however. He did remember that it had been an octagonal one, which tallied with the facts. He had no idea that he would cause the man so much trouble, but now that it had happened he was not sorry, because it would be important to science that we had these experiences of a developed person at our disposal.

I was very interested in finding out whether our friend was still convinced that an object in Spaceland cannot be turned around in such a way that left and right become interchanged—but I was a little reluctant to bring up this touchy subject. Whether the Sphere sensed that I was thinking about it I do not know, but while I was searching for words to continue the conversation, his thoughts had been occupied with the same subject, because he said suddenly: "Mr. Vertato's experiences are entirely analogous to such experiences in Spaceland."

"Do you mean that people were also turned around there as the result of a swing through an area of the fourth dimension?" I brought out. "But isn't that possible only through the intervention of a four-dimensional creature?"

"Yes," the Sphere said. "I talked this question over with the Over-Sphere. He convinced me that it is indeed possible to turn three-dimensional creatures around. Since he did not want to start with living beings, he conducted his first tests with objects. Well, they do not reveal much, but then I conceived the idea of turning books around. With this in mind I took the Over-Sphere to a bookshop where I pointed various political books out to him—all of a political tenor with which I was in complete disagreement. He turned them around. The print changed into mirror script and became completely illegible. It had not occurred to me, however, that I would dupe no one but the bookdealer, who was totally innocent. The books could not be sold. The bookdealer returned them to the publisher who either replaced them with other copies, thereby incurring the loss himself, or said that he did not know these books and would have nothing to do with them. One or two copies landed in a museum where they were exhibited as an amazing but inexplicable freak of technology, along with an unreversed sample of the same work for comparison. All I had really accomplished was to draw more attention to the books themselves rather than limiting their distribution, which is what I had wanted to do in the first place."

This took a while to digest. After a long silence I asked: "Have living creatures also been turned around in Spaceland?"

"Yes, that did happen. At first, the Over-Sphere did not want to hear of it because of the possibility the victim might suffer from it or become unhappy as a result. And at first I did not understand this. How can anyone be made unhappy by such a reversal? I wondered. But the Over-Sphere persisted and so I had to think of something that would prevent the subject from *becoming* unhappy or simply *feeling* unhappy as a consequence of the experiment. While I was thinking about it, a unique plan occurred to me. There is a strange disease in Spaceland called 'left-handedness.' Persons suffering from this malady perform everything from the left that others do from the right, with the result that they are handicapped in daily life where many actions have to be carried out from the right and someone who is a lefty cannot do them or else does them poorly. Other actions which can be performed as easily from the left as from the right by most people are done from the right only because they are right-handed. If a lefty fails to do this he is laughed at by the others."

"Just because it is not customary?" my grandson asked.

"Actually, yes," the Sphere answered. "Fashion is king in Spaceland just as it is in Flatland."

"It is probably the same in Lineland too," my son added.

"But what did those lefties have to do with your plan?" I wanted to know.

"Well, that is rather obvious. When a left-handed individual is turned around through the overspace, left and right are transposed for him and he is accordingly changed into a right-handed person."

"A wonderful plan," I exclaimed. "And did it work?"

"Yes," the Sphere said. "Or rather, not really. On my instructions the Over-Sphere turned one such individual around."

"So it did work," my son said.

"That is what we thought at first, but even though this person was now right-handed, he was very unhappy. You see, he had learned to write on the left side and when he now did this, a writing would result which he could read himself but which was illegible to everyone else. It was mirror writing. And while he could read his own script, he could not read that of others. Even the letters he wrote just before his cure now looked like hieroglyphics to him. It was so bad that it actually made him sick. The doctor couldn't find anything wrong. He told him to stop reading and writing for the time being and to take long walks, but this turned out to be fatal. Now that left and right had become interchanged for him, he moved out of the way of an oncoming vehicle on the wrong side. He died of the consequences."

"That was a tragic ending to a well-intended experiment," I remarked.

"Yes," my son added. "The patient was released too quickly. He should first have had a chance to become thoroughly accustomed to his changed circumstances. Wouldn't it be logical to keep such patients, in the future, in a specially designed rest home where they would be taught gradually how to act in a world that has become reversed for them?"

"And music might well play as important a role there as in the case of Vertato," my wife added.

But in general we were convinced that the cure was not significant enough to be repeated on a larger scale. The problems of a left-handed individual in the world are not so extensive that he no longer feels at home in it, and a sudden change seems to involve so many dangerous consequences that the disadvantages are greater than the advantages.

The Sphere now took sudden leave and disappeared, leaving us behind with much material to ponder and discuss.

15 A Rumor

Daily life went on placidly and the mysteries of space, which had caused such a commotion at first, now began to fade more and more into the background. The average person, who did not really understand much about it in the first place, had already long since become interested in other things and might at most think to himself: "You can never tell what those scientists will think of next!" The scientists themselves, and here I am speaking of the mathematicians and physicists, were now convinced of the possibility that worlds with more than two dimensions did exist, even though most still doubted the *fact* of that existence. While they regarded the stories about a visit from a so-called Sphere or Over-Circle from a three-dimensional world as pure inventions, they also attached some value to them, seeing that they could be used to familiarize a larger public with the geometric characteristics of other conceivable worlds. But amazing things were to occur which, very shortly, would throw a new light on the matter.

I still remember vividly that we were all gathered in a family circle one evening, comfortably chatting of this and that, when one of my children asked whether we had heard that the director of the Trigonometric Service had been discharged. I was stunned, because even though I did not know this man personally, I knew that he was a conscientious civil servant, proficient in mathematics and technology and known as a meticulous observer. What could be the reason for his sudden dishonorable discharge?

My wife suggested that he might have become involved in some one-sided financial dealings, something which does occur at times among high-ranking civil servants, but I could not see how a man to whom science was so important and who furthermore enjoyed a handsome salary could succumb to flagrant dishonesty.

"Of course we don't know," my wife replied, "what kind of financial problems he may have had. For all we know, he may have lost a great deal of money on the stock market."

"Or has a spendthrift wife," my son observed.

"Or a real rascal for a son," my wife countered.

Meanwhile, we still did not know the real reason and it all remained mere speculation.

"What does 'Trigonometric Service' mean?" my oldest grandson wanted to know.

"Triangle-measuring Service," I explained.

"Yes, I know that," the boy answered, "but, after all, everyone is a measurer of angles. When I meet someone outside, I measure the angle he has turned toward me simply by estimating it, and we all know the degree of accuracy we have achieved in this, probably as the result of native experience. At school we were made to practice this angle-measuring business ad nauseam. And if one is able to measures angles, one naturally learns to know triangles."

"And yet," I interrupted, "this serves a real purpose.

Time spent on it is not lost. Someone who has learned to estimate angles quickly and accurately will at the same time have acquired considerable knowledge of people."

"I don't deny it," the boy resumed, "I was only talking about the concept of goniometry and I maintained that it is neither something special nor very difficult. If we want to determine angles more carefully, we use the touch system: why, one can even have his angles determined scientifically at certain special institutes which provide the individual with a certificate of the results. I never heard the term 'Trigonometric Service' used in this connection."

"Very true," I declared. "But the Trigonometric Service is something else — it is an institute which was established for the purpose of mapping the world accurately by means of trigonometry. The entire world is divided into triangles for this. The angles are measured and the length of the sides calculated."

"But, Grandfather," my grandson cried out, "you know better than that! How can the length of the sides be calculated when only the angles are known?"

"No," I answered immediately, "you are right. Naturally, one side has to be known. Then, if the angles of the triangle are known accurately, the other sides can be calculated."

"Only one side and two angles have to be known. The third angle can be calculated easily because the three angles of a triangle together equal two right angles or 180°," the little squirt thought it necessary to point out. All the same, I was proud of his observation. It showed once again how closely this little fellow was following the tracks his forefathers had made. Wasn't our ancestor, the famous Square, his great-great-grandfather?

Silence prevailed for a while. Everyone was mulling the case over. My wife was the first to resume the conversation, saying: "I still do not understand how such a position as director of the Trigonometric Service can lead to any sort of fraud. All it involves is measuring angles as accurately as possible and working out the calculations. The director might have to decide which points should be considered as corners for his triangular net, but it's still simple technical survey work."

"We don't know," my daughter-in-law interposed, "what sort of a relationship this man had with his subordinates, and to what extent some dishonesty might have entered in there."

"Let's not go off on wild tangents," I admonished. "We are merely speculating. We know nothing at all about it and there is no point in letting our imaginations run away with us."

"We might hear more about it in the near future," my wife said.

"Quite likely," I said, not suspecting that this would really prove to be the case.

16 The Visit

One evening a few days later, I sat in my room studying. My favorite subject, mathematics, had seized me again and I was immersed in complicated problems when a visitor was announced. I had him enter. He was a stately octagon who moved with grace and inspired confidence and respect from the first. I was greatly surprised when he introduced himself as Mr. Puncto, the dismissed director of the Trigonometric Service, who had been the subject of so much commotion.

"Perhaps you have some idea about the purpose of my visit," he said.

I did not answer him right away. Some thoughts were indeed running through my mind, but it did not seem proper to voice them. I thought that he might want to ask for my help in connection with his dismissal. A man, even an intelligent and well-educated man, can be at his wit's end and driven to resort to a soothsayer or quack—didn't I have the reputation among the general public of someone who practiced black magic? Didn't they regard me as one who possessed secret powers? It's true that my visitor did not seem to be the sort of person who would credit that sort of talk, but in a desperate situation, and perhaps at his wife's urging, a man might well seek such a recourse as his final hope when all other means had failed him.

What did he want of me? Might he have to produce a large sum of money, which he had either misappropriated or lost, and did he think I would perhaps be able to conjure it up for him out of the three-dimensional world? These things were going through my mind and I did not know how to answer my visitor. He must have noticed my confusion because he continued after a few minutes, saying: "Briefly, the matter is as follows: I have come to ask your advice in connection with my dismissal from the Trigonometric Service, of which you have no doubt heard.

"I should add," he went on, "that the wildest rumors have been circulating about this affair, but I believe I am right in thinking that your view will differ from the rest."

"I must admit," I answered, "that I only know the fact of your dismissal and am altogether in the dark about the reason for it. If you want to tell me more about it, I will be able to determine my own attitude more clearly, but I really don't believe that I can be of any help to you."

If my words had a slightly hostile ring, I meant them to have it. I was resolved that the moment my visitor asked for my cooperation as a magician, I would pointedly show him the door—in spite of his eight regular sides.

"I understand," my guest said, "that you do not see yet in what way you can help me, and I cannot tell you either, because I also don't know. I am still living in a complex of problems whose solution is far away and obscure to me. I am still in the dark about the very direction in which to look for an answer. But if *you* cannot do it, no one can. There are enough mathematicians in the world, in our two-dimensional world," he added with a wink, "but it seems to me that we will never get out of the woods with our ancient, classical form of mathematics. And because you have shown yourself to be receptive, not only to new concepts, but also to the propagation of new ideas—provided they seem to you to justify this—I hope to find you able to view the facts with an open mind."

"I am beginning to see," I said, "that the entire question is connected with a mathematical problem. In other words, it does not involve . . . a legal or financial matter?" I ventured to ask. "You will forgive me if I ask you this quite frankly."

"Of course," he answered. "I am, on the contrary, very happy to be able to dismiss all the talk invented by the public, which is unable to understand even the smallest part of the problem involved. It is a purely scientific, at least technical, question—but . . . an extremely strange one," he added.

Not only did my earlier antipathy toward him now fade away, but my interest in the problem had been thoroughly aroused—even though I knew nothing about it yet. And so I said: "May I ask, does this involve some difficulty with the application of measuring methods or with the calculation of mistakes that occur during the measuring process? You must bear in mind that I am not a technician."

"As far as I can see, it is not a technical, but a theoretical problem," he replied. "Though I am not certain. It is all so strange, so very strange!" And with that he appeared to be gazing off at faraway vistas. I thought he was grieving, that he felt himself to be confronted by an insoluble mystery, a mystery with which others could not even sympathize. And I thought to help him with the observation: "Might it not be better to leave things as they are? I mean, if there is really no answer to the problem . . ."

But this made him flare up: "No solution to the problem? There *has* to be one. Everything has its reason. There is an explanation for everything. That is why I thought that you, as a mathematician, would agree. The explanation might be strange . . . even very strange, perhaps!"

"My dear sir," I said, "you can count on me. In any case, I want to go into the problem thoroughly. Can you describe the essentials to me in a few words?"

"It's better that I don't," he said. "You might question my observations or even my mental powers. Do you have the time and opportunity to accompany me tomorrow?"

"Yes," I said, "I have no objection—on the contrary, I should like very much to get closer to this question which has already made me very curious. But I can see it will be a big disappointment if I fail to find an explanation for the phenomenon which is causing you such problems."

"In the first place," he argued, "you might catch me in an error, though I frankly doubt it. But if you do, the matter will then be easily cleared up, and I shall be extremely grateful to you for your cooperation. Then again, my observations could turn out to be correct, and as inexplicable to you as to me. Then I would no longer stand alone but we could be sharing the difficulties and could discuss them. That would be a great comfort to me. Thirdly, I do not consider it impossible that you will find a solution to the puzzle. Will you help me?"

"As I said," was my answer, "from now on I shall take great interest in the matter. I will be most happy to do everything in my power, but the outcome is doubtful and I hope that you will not want to reproach me with anything afterward."

"Reproach?" he said. "Never. Who knows, perhaps we'll find the solution together. If two minds can discuss a problem there is always a chance that a solution will be found—but it will be very difficult."

"Can I learn a little more?" I asked when we took leave of each other. "Couldn't you lift a tip of the veil for me, so that I can know in what direction to look for the problem?"

"All right," he said. "Tell me what the sum of the angles of a triangle is."

That unexpected question did surprise me, I must admit, but I answered: "180°, of course."

"Always?" he asked, and he left.

17 Amazing Results

This last comment of the former director gave me food for thought. I had been talking with him for quite a while and thought I was dealing with an intelligent man, but after this question, this doubt of his . . . was the man really in his right mind? Was he all there? Whoever doubted that the sum of the angles of a triangle is 180°? Who has ever seen any other kind of triangle? Or can conceive of such a triangle? Besides, it is easy to prove that the angles total 180°. Hadn't I better withdraw from the case? It would be difficult to get rid of the man later, and just try to talk a maniac out of some senseless truth or other once he has it in his head!

These reflections caused me a sleepless night. I considered inventing some sort of excuse, an illness, an unexpected summons—if need be, a death in the family—in order to get out of it, but when he stood in front of my door early the next morning, he inspired me again with such confidence that I decided to go along with him, and to the present day I have never regretted it.

He brought me quite a way out of the city. Here a three-cornered net had been set out with observation posts at considerable distances from each other. We made our way to post *A*. By means of an accurate trigonometric instrument it was possible to determine the angles between the directions *AB*, *AC*, *AD*, *AE*, and *AF*. My guide asked me to try using the instrument myself and measure the angle between the direction to *C* and the one to *D*. I found myself enjoying it and even measured the four other angles.

"Together these five values will naturally have a sum of 360°," I declared.

"Let us try," responded my new friend— for why shouldn't I call him that—"to put it to the test."

A net of triangles is laid out.

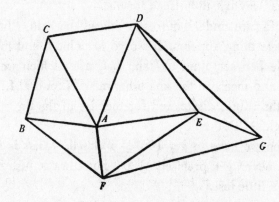

This we did and the result did turn out to be a pretty accurate 360°, but not *precisely* 360°!

"Of course one will never get exactly 360°," my guide said. "Every observation that is carried out, with the utmost accuracy even, entails a small error, an observational error. Because of this the final sum will never be quite correct, but a little too much or too little. The trained observer will get a smaller deviation than an untrained man, but even someone working with the closest accuracy possible will have a small error in his result."

"You don't need to tell me that," I said, "I have long been familiar with observational errors."

"Then you will also understand," the other resumed, "that with the successive repetition of the same measurements the result will sometimes be too little and sometimes too much."

"Of course," I said, "but one can also have a result that is too much, several times in succession."

"Or, on the other hand, one that is too small on successive occasions," my friend finished for me.

"So far we understand each other," he said, "but now the strange thing appears. If we go to point *C* and measure the angle between lines *CD* and *CA*, after which we go to point *D* and measure the angle between *DC* and *DA*, finally adding the values discovered for the size of the three angles of triangle ACD . . ."

"Then we have to get 180°," I stated. "That is to say, we will never get precisely 180° but always just a little more or a little less."

"Exactly, that *should* be so," he said, "but it is *not* the case. I am always getting too large a result and the deviation I find is just a little too great to be attributable to observational error."

"This means," I concluded, "that the deviation cannot be the result of observational error, but that it is a true deviation, that therefore the sum of the angles of this triangle is really greater than 180°."

My conclusion was right, it could not be refuted logically, but when I had said it, I was shocked by my own words. For how could that be? What had I just said? The sum of the angles of a triangle greater than 180°! That is impossible! Just how could that be?

My travel companion noticed my confusion. He smiled and said, "Yes, that is right, in my opinion that is the only correct deduction. But how is it possible? After all, we can't claim that we, for the first time, have discovered a triangle the sum of whose angles is greater than 180°. That would be contrary to reasonable intelligence. It runs counter to the very foundations of geometry! And the measurement holds not only for this particular triangle but for all triangles of the trigonometric net. I just can't find my way out. Can you help me? Can you make a guess at the underlying reason for this amazing result? Could you make even a small suggestion about where we should look for a solution?"

"What really did happen?" I asked. "Was this the reason for your dismissal?"

"Yes," he said, "this is the reason and no other. They simply refused to accept the result. I was called up to testify, they heard me out, I saw that they looked at me with pity. Either I was a bungler or I was not entirely in my right mind. The sum of the angles of a triangle *is* 180° and that's that. But by the Holy Circle, I know that too! In their opinion, someone who dares to consider such results is just not fit for my position."

"But, after all, one cannot change the data one has," I thought out loud.

"It pleases me to hear you say that," he remarked. "But I had not really expected you to say anything else. Look, it would be fairly easy for me to tamper with my measurements, canceling out the discovered difference, but I couldn't answer to myself for that."

"Besides," I continued, "others might conceivably discover the difference at a later date and bring your cheating to light."

"That's certainly possible," he said. "No, I really cannot do otherwise."

"There is one other way to get out of the muddle."

"And that is?" he asked.

"Well, simply this, that you record your results with a little less accuracy, rounding them off, in other words. The small differences will then drop out and the results will check out."

"That is a possibility, but it would go against the grain with me. I do my work the best I can and I give the results of my measurements with the greatest accuracy possible. I cannot do otherwise! I consider that the only proper attitude toward my work."

"I understand you," I reassured him. "I would look at my work in exactly the same way."

A long silence followed; we were both busy with our own thoughts. I was the first to speak.

"Let us consider what we have to do first of all. In my opinion, we must establish this discovered phenomenon as accurately as possible."

Full of enthusiasm, he replied: "My dearest friend, it does not happen often in life that, in later life, one suddenly acquires a true friend. Most friends are made in youth, but now I feel that I have found an extremely close friend in you. Please let me call you that. My dearest friend, I cannot tell you how grateful I am for your offer to help me, for your readiness to give time and thought to help solve a difficulty which concerns me personally."

"The problem does not concern you alone," I said. "It is a scientific problem in the field of geometry and it fascinates me. I shall be delighted to try and find the solution together with you, as two trusted friends—or, at least, if a solution is not to be found, to work out the problem in words and figures as clearly and briefly as possible so that someone else will be able to use our results at a later date."

"That is the true standpoint of all science," he said. "Come on, let's get to work!"

"First of all," I said, "why do you think this deviation was not discovered until now? Was the observation method in the past perhaps not as accurate as it is today? Or is there another reason?"

"That's easy to answer. Due to more improved instruments, the method is much more accurate than it was formerly, but that does not alter the fact that the deviations are so great they should certainly have been discovered long before this."

"In your opinion, why didn't they discover this earlier? Could it also have involved the fear of making a fantastic, inexplicable result public?"

"Your explanation is possible, of course," Mr. Puncto replied, "although it would discredit earlier observers and we should not do this until there is absolutely no other choice."

"Then let us first look for another reason," I continued. "Did the former observers give rounded-off results or were their results worked out to as many places as possible?"

"I looked into that and discovered that they gave their results as accurately as possible. However, there was never any question of too great a value of the sum of a triangle's three angles. The small deviations in their results can be ascribed to observational errors."

"In that case we ought to investigate whether there were any differences between the old observation method and the present one. In short, what *is* different and what is *done* differently?"

"Well," he said, "there is a difference in one aspect. Formerly a net of small triangles was used, and now we are using very large ones."

"Then we will have to determine whether the size of the selected triangle has any influence on the result. How can we best do this? You have been dismissed. Does that mean you are no longer able to carry out the observation work because you no longer have the necessary instruments and assistants—and, if so, can we get the instruments and trained helpers in some other way?"

"Well," he sighed, "I believe that my men, once I have explained the matter to them, will be perfectly willing to make all the measurements I think are needed."

"Fine," I said. "I hope to see you at my home as soon as you have looked into this and then we will be able to talk about it further."

We parted the best of friends. I was convinced that Mr. Puncto would make his investigation with the greatest accuracy and speed—and in this I was not deceived.

18 An Impossible Problem

Scarcely a week had passed when my friend Mr. Puncto stopped by to discuss the results of his investigation with me. He had laid out a great many large and small triangles and measured their angles with the utmost care. The result was that the sum of the angles in a triangle consistently exceeded 180° by a certain amount, an amount which increased with increase in size of the triangles and could not be measured for small triangles.

I spent many hours checking his calculations but could not find a single error.

"Well, these facts are now established," I said emphatically. "At least, we're one big step further."

"I don't agree," Mr. Puncto rejoined. "The case remains equally curious and we don't have the slightest glimmer of an explanation."

"Things are not that bad," I reassured him. "First the facts, then the explanation. We're well on the way and I have no doubt an explanation can be found."

We sat up late at night theorizing, but without success. It really was a curious business! Finally we had to break our meeting up, but we agreed to continue giving this a great deal of thought. Perhaps one of us would come up with a fine idea. We planned to hold our next meeting three days later and share our thoughts with each other—even those we ourselves had rejected as worthless. Sometimes a wrong notion can lead someone else to a workable thought. We said a warm goodbye, fully convinced that together we would find our way out of the labyrinth.

I need not tell you that I was constantly preoccupied with the problem. Even when I went to sleep I would continue thinking about it as long as possible in the hope that something might come to me in a dream, which even if it would not be the solution would at least push me in the right direction. All in vain, however. During the day I reasoned out loud with myself, trying to set up a logical argument, but it always bogged down somewhere, since an absurd question indubitably must fail to produce a logical answer. I became impervious to my surroundings and saw only triangles and more triangles in front of me, big ones, small ones, isosceles, equilateral, irregular, all sorts of triangles. In my thoughts

I walked around them, measured the angles, and then thought immediately of the next triangle.

My family left me alone because they did not want to disturb me. From time to time they would look at me questioningly. Whenever someone asked me something, I would not hear him but respond instead with a peculiar counterquestion, such as "Show me a triangle the sum of whose angles is more than 180°."

On the morning of the third day—the meeting with my friend Mr. Puncto was to take place later that evening—my little grandson entered my room. "Grandfather," he said, "I don't know whether I can help you or not, but all day yesterday you were asking for a triangle the sum of whose angles is more than 180°."

"Yes," I said. "I know what you must be thinking, my boy. Of course you think that your old grandfather is beginning to lose his mind. And it looks a little that way, I know, but I also know that I'm not. I am looking for the solution to a problem and I am not even sure whether it is a mathematical or a philosophical problem, but in order to solve it, I must be able to visualize a triangle the sum of whose angles is more than 180°."

"Yes, Grandfather," the boy said, "I know that an ordinary triangle cannot be different from any other triangle that you can imagine. Naturally it has to be a peculiar triangle, because only a peculiar triangle can have a peculiar property. Shall I show you a triangle like that?"

Even though I had no confidence, something more than

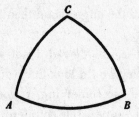

My grandson shows me a triangle, the sum of whose angles is greater than 180°.

love for my grandson made me answer that I would certainly like to know what kind of a peculiar triangle he had dreamed up. The boy was visibly pleased that his grandfather showed so much interest in his discovery and he began right away to lay out a figure.

"Here you have a triangle, Grandfather, the sum of whose angles is larger than two right angles."

"Yes," I objected, "but the sides of that triangle are not straight."

"That's just the point," the young virtuoso said. "That is my discovery. After all, I had to think of something peculiar in order to answer a peculiar question."

"You're right," I told him. "But what shall I say? The triangles I saw did not have curved sides."

"Are you sure of that?" the boy asked.

This irritated me a little. Of course I was sure. While the sides of the triangles we had measured outside had not been drawn, they were light rays coming from other corners to our observation instrument, and if light does not travel along straight lines, then nothing makes any sense!

Nevertheless, I was touched that he had wanted to help me out, and I didn't want to show my annoyance. Even though

he hadn't brought me one step further, he had at least interrupted my gloomy train of thought, my hopelessly confused
speculations which constantly turned in the same direction
and could not give me an answer. In the evening my friend
would come to talk with me about this strange case, whose
solution now seemed very far off. But then—perhaps my
partner had been able to find an answer to the difficult
problem after all, or at least discovered *something* that
would guide us in the right direction.

19 *Strange Triangles*

My hopes that Mr. Puncto might have had a brain wave
proved to be vain. He had been walking around with the
problem for some three days now, as he explained, but I
understood from his words that he depended entirely on me
for the solution. He believed that as grandson of the famous
Square, I was bound to have unusual gifts which could be
applied successfully to the most unusual problems, especially
those for which ordinary mathematics failed to provide even
the glimmer of an answer. This was flattering, but still it
irked me that he had not really given it his best effort. I told
him this but he laughed and bounced the reproach right back
at me when I admitted not having the slightest idea what
road to take next. I could not contribute anything that would
propel us in the right direction, a direction which would
bring us just a little closer to our goal.

"I am not even asking for that," my guest said,

"whether the direction is right or wrong will become clear only later. For the time being, we will have to accept any and all extraordinary notions leading us to triangles with other properties than we are used to, as a possible step to a solution. If I could only think of a figure which I could pass for a triangle and which has the property that the sum of its angles is greater than 180°, I would regard it as a way out of our labyrinth; but so far I have not succeeded in coming up with such a figure."

"Yes," I had to acknowledge, "that is one point of view. If you put it that way, I might be able to come up with something." And I produced the triangle with curved sides which my grandson had shown me.

Mr. Puncto listened attentively, examined the triangle in question very carefully, and did not laugh. Finally he said: "Perhaps this is a first step in the right direction. It is a solution to the problem insofar as the sum of the angles exceeds 180°—but, on the other hand, I wonder whether we can really accept a triangle with curved sides."

"Of course we can't," I snapped a little impatiently. "Light rays move along straight lines."

"That is true, of course," he answered calmly. "And I can therefore consider this at most as a first step, and possibly even a step in the wrong direction."

We continued to discuss our problem a long time, for many hours in fact, without getting any further. When we finally parted, my friend said: "Let's quickly summarize the way it looks. Through measurements we have found that

the sum of the angles of our triangles does not equal 180°
but is greater, and that the size of the difference seems to
depend on the size of the triangle. Larger triangles have a
greater deviation than smaller ones. Secondly, we can only
conceive of such a deviation occurring in triangles whose
lines are not straight lines."

"That's the dismal upshot," I said.

"But it is only temporary," he declared optimistically.

"Yes, but what next?" I wondered. "Do we have to walk
around another three days waiting for another happy inspira-
tion? A better one, I hope?"

"I don't see much future in that," he replied. "It might
be better if we were to talk a little with others and see how
they react to these ideas."

"Perhaps I ought to consult my grandson again," I
suggested a little sarcastically. "It looks to me like the right
way to turn him into a conceited little prig."

"I would rather find out," my friend stated, "how men
of science and especially mathematicians react to this.
Couldn't we stir up some interest there? We now possess
observation data which warrant a scientific explanation."

"Or refutation," I added.

"They cannot be refuted," he said, and I had to agree,
because the observed facts were undeniable.

I also thought it best to call on the resources of the
Mathematical and Physical Sciences faculty, and I undertook
to try and interest these people in taking cognizance of the
problem.

20 The Faculty

I succeeded in doing this above every expectation. I had
thought that the faculty with whom I had requested an
audience would appoint a single professor to hear our
"defense" of the amazing events—but it turned out differently. Mr. Puncto and I were invited to attend a plenary
faculty conference where we would be able to present our
views.

In high spirits we made our way at the appointed time
to this solemn session, but the moment we were admitted we
got a strong feeling that they had summoned us here to put
an end to our "nonsense" once and for all, through an official
verdict issued by this high-ranking college of experts—and
this feeling stayed with us throughout the session. The chairman first turned the meeting over to Mr. Puncto, who was
addressed with the title "former surveyor." He was quick-
witted enough to change this somewhat mockingly into
"former chief land surveyor, ex-director of the Central
Trigonometric Service." But the chairman simply reacted to
this with a curt "You have the floor," whereupon Mr. Puncto,
calm and businesslike, proceeded to set forth the entire case.
He related how the measurements made under his guidance
showed that the sum of the angles of the triangles in question
was greater than $180°$, increasing proportionately with the
increase in the size of the triangles. The reason this had never
been discovered before he attributed to the fact that formerly
only relatively small triangles had been used and there the
deviation was smaller than the observational error. He

asserted that a series of experiments ought not to be rejected simply because their result is found to be strange, but that attempts should be made to explain such a result scientifically.

After he had explained all this clearly and succinctly, the professor of mathematics, Ergo, asked for the floor. In a long-winded argument he contended that it is undoubtedly true that science must try to explain observed phenomena, but that these facts in question were of such a nature that a scientific man would have to reject them at once. The angles of a triangle together are 180° or two right angles. This has always been so. It is an undeniable fact and can never be otherwise. If a series of observations is in conflict with the first fundamentals of mathematics, the series is false. It is not up to scientists to track the error down—that has to be done by those investigators responsible for the mess in the first place. The faculty should put the matter aside. It would be beneath their dignity to involve themselves even in the slightest degree.

After this, the professor of physics, Professor Supposo, spoke.

We felt right away that here was a different man. He built up an argument to the effect that the science of physics certainly had had to cope more than once with results which at first seemed very strange, but which, upon closer inspection, had turned out to be correct. Now it was true, he asserted, that these facts were extremely strange, because the very first principles of mathematics were thereby attacked, "but," he said, "we will nevertheless treat these observations with the greatest goodwill. We must ask ourselves: Can we conceive of a triangle the sum of whose angles is greater than 180°? We have never yet seen such a triangle, neither in reality nor in our thoughts. It will therefore have to be a very strange sort of triangle which has such a strange property. Can either of the gentlemen who have come here to throw over the existing and dependable science name such a triangle or show it to us? If not, then I consider the matter closed. But if they can do it, I am prepared to think the matter over further."

Fortified by this accommodating trend of thought, I felt myself called upon to present my grandson's hypothesis. I therefore asked for the floor, and after getting it I spoke as follows: "Gentlemen, very learned gentlemen of this illustrious college, it is my privilege to make a few comments on the words of the previous speaker. I am in a position to show you a triangle which satisfies the required property. The previous speaker has already established that it will have to be a strange, an unusual triangle. Well then, I tell you that a triangle, the sides of which are not straight but curved lines can have angles the sum of which is more than 180°."

There was a moment's silence of which I made use to delineate such a triangle.

"May I point out here," Professor Ergo said, "that in the world, light travels in a straight line and that the sides of the triangles used by the trigonometric measurements therefore cannot be curved."

A murmur of approval died away when Professor Supposo replied: "Mr. Chairman, I am not in complete

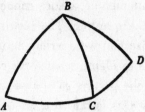

Side BC cannot be curved simultaneously to the left and to the right.

agreement with this. If it should be necessary to suppose that light does not travel in a straight line but in a curved line, then investigations would have to be made whether this hypothesis is not in conflict with other phenomena—but I have an entirely different objection to the hypothesis. Let us consider two joining triangles of the trigonometric net, *ABC* and *BCD*. Because the sum of the angles of *ABC* must be greater than 180°, the side *BC* now will be curved outward, or from here to the right. But triangle *BCD* will also have to manifest the same phenomenon. The sum of the angles of this triangle, too, must be greater than 180°, and this requires that the side *BC* be curved to the left. It is clear that the line *BC* cannot curve simultaneously left and right, and this breaks down the entire explanation which our guest gave."

I could not answer this. Professor Supposo's argument looked so airtight to me that my hypothesis would have to be canceled out. There was nothing left to say because my friend Mr. Puncto did not have anything else to contribute.

In a sarcastically friendly tone the chairman said: "You see, gentlemen, the faculty has spent every effort and considerable time on your problem, with no result. Now it is up to you to find the errors in your observations—certainly these gentlemen of science cannot be expected to spend more time on the question. The faculty asks of you only that you will realize you were wrong and that you will not try behind our backs to blame science for allegedly lacking in cooperation, thereby putting us in a bad light. If it should ever appear to us that this is the case we will not hesitate to inform the law and to sue you for defamation of character. Your presence among us is no longer appreciated."

Neither of us dared to reply. We left the hall and with heavy spirits walked toward my home. We had lost the battle, but we were still not convinced that we were wrong. On the contrary, the facts were in our favor and it was up to science to explain the phenomena. I asked my friend to come home with me to discuss the matter further.

Once we were back in my study we could relax again, the hostile world was kept outside the boundaries of the room. For a long time we remained sunk in our thoughts until Mr. Puncto broke the silence. His words reflected an indestructible spirit of optimism.

"In spite of all this," he said, "the interview with the learned gentlemen has not been unprofitable. Professor Ergo, a gentleman of the old school, simply rejected what he did not understand, but Professor Supposo said something more, and his argument included some things worth considering. He wanted to see a triangle which satisfied the strange characteristics, and—if need be—he would be satisfied with a triangle with curved sides—only the direction in which those sides curve involves an impossibility. It is true that I do not see that we have come one step further, because his conclusion seems perfectly correct to me. If a line cannot be curved to the left or to the right, it cannot be curved at all, because there is no other direction in which it can be curved."

"Not to the left, nor to the right, there is no other direction in which it can be curved," I repeated after him. "There is no other direction in which it can be curved. Come now, we have found ourselves thinking so often that there is no other direction! We might not be able to see it, but a direction certainly does exist which we cannot perceive, a direction perpendicular to our world. The lines appear to be straight to us, but perhaps they are curved in that invisible direction. After all, that can just *be* the case! That *must* be the case!"

"I don't understand very well what you are saying," Mr. Puncto resumed, "but I do understand that you see a solution."

" 'See' is not the right word," I said, "not 'see,' because I cannot see it, but understand, 'understand' is perhaps right. Let me think about it a little more and let us continue our talk tomorrow."

And with renewed courage we said goodbye.

21 Vision of Circleland

Would nighttime bring the solution or at least guide my thoughts in the right direction? For how often has sleep turned out to be favorable to finding the right thought pattern, already started much earlier, sometime during the day perhaps. In sleep at night, when all disturbing factors have been switched off, such as caused by outside noises and occasionally by faulty thought combinations, a logical thought pattern often appears which was not to be found in the waking hours.

I had faith in the night. I understood, I felt, that I was so close to the right conclusion that my mind could make the final step in a dream. I would simply have to order my thoughts well beforehand, in line with the principal points. How did it go again? What I had to see was a triangle, a triangle with three angles which together exceeded 180°. The three sides therefore cannot be straight. They have to be curved, curved, but not to the left and not to the right, but still curved. I just do not see it! I cannot see it! They are curved in an invisible direction. But how is that possible? Curved in an invisible direction! How? Is an invisible direction conceivable? Yes, of course, because while we have two directions, the Sphere comes from another one altogether, one which we cannot see, but which is visible to him, a direction perpendicular to our world. Could it possibly be that the side of the triangle is curved in that third, invisible direction?

We ourselves can never see such a curvature, but we can observe it in the world of one dimension, Lineland. Can there be a curvature in Lineland which is invisible for the Linelanders, but visible to us?

And I saw Lineland before me, but Lineland was not straight, it was curved, curved into a very large circle. It isn't Lineland, it is Circleland. Look there, the inhabitants are all moving back and forth without being able to pass

Circleland.

each other. How self-assured they are, humming back and forth. Ah, there is the king!

"King," I said, "Sire, please listen to me for a moment!"

"To whom should I listen?" the king cried out. "I don't see anyone. Is it that wicked magician again who was endangering my kingdom years ago and who vanished into nothingness just in time? Do not let him come into my sight again! I'll . . . !"

"Sire," I said, "I am that 'wicked magician'! But I am not wicked and no magician! I come from the world of two dimensions."

"What are you talking about, spirit or spook, or whatever you are? A world of two dimensions? There is only one world and it has only one dimension. That is my world and I am its king and I will not tolerate any argument, not from my subjects and certainly not from others, whether they be spirits, spooks, or magicians."

"Now listen here, Big Chief," I said.

"That at least is respectful language," he declared. "I am the chief of this world and of course I am big and important.

What do you want of me? Did you come to continue your talk about another world, the one that you come from? Nonsense. You can't make me believe that!"

"Actually," I told him "I would prefer to talk about your own world this time."

"All right," said the king, "I see that I will just have to be very explicit to make clear to you what my world looks like. That I know best, and I can see that your ideas about it are pretty confused."

"I am listening; do go on."

"Well, it is actually very simple—at least for someone with brains," he added disparagingly.

"I am listening patiently," I said meekly.

"My world," said the king, "*the* world, because no other is conceivable, for where could that be? Nowhere . . ."

"Where then am I?" I ventured to interrupt. "Aren't I outside of your world?"

"Where you are, and whether you are anywhere, I just don't know," said the king. "Probably you do not exist at all and are simply a dream vision of mine."

"But don't you remember," I continued, "that I punched you in your insides? And wouldn't you say that is possible only if I am outside your world? Shall I do it again?"

"Nonsense," said the king, "a stomach cramp can come any time and is not the result of a punch coming from outside the world. Nonsense!"

"All right then, why don't you start with explaining what sort of form your world has?"

"Form! Form! What is form? A creature can be longer or shorter. Is that a difference in form? And all creatures are situated right behind each other in an infinitely vast world, because the world is infinitely vast. From both sides the world stretches out into infinity."

"Allow me one single question, Mr. King. How do you know that your world is infinitely big?"

"Funny question! How could it be otherwise? If the world stopped somewhere, what would be beyond that? After all, that line can be extended continually, continually, past all borders. And that is what is called 'infinite.' "

"But do you ever get any reports from creatures who are located infinitely far away? Even if it is only a single sign of life?"

"No, of course I don't, and if you had any brains at all, or could at least think a little, think logically, you would understand that. Even if the world extends into the infinite, it still does not have to be populated all the way into infinity. But suppose that *is* the case, then we would still not be able to receive any signs of life from infinitely distant residents because their cry would reach us only after an infinite length of time. Secondly, the volume of sound diminishes gradually with distance. Beyond a certain limit, we are no longer able to hear any sounds. If you possess any intelligence whatsoever, this explanation should suffice to give you some notion of the structure of the world. Do you have any other questions, stranger?"

"No questions," was my answer, "but I do have something to say to you. In the first place, I want to point out that your world is not straight, but curved; furthermore, it is not infinite, but finite. It is what we Flatlanders call a circle, and not a straight line."

"But," the king interrupted, "how can the world not be infinite? If it is limited, what is behind that limit?

"Your world is not infinite," I said, "but it is also not limited. It forms a closed line which returns into itself. If you could move many miles forward in the same direction— which is impossible because your fellow citizens are in your way—you would return to the selfsame spot where you are now."

"I don't understand that at all," said the king.

"No, obviously not," I resumed. "I don't blame you. Your world is finite because it is a circle, a curved line which is closed."

"But how can our world be curved?" the king asked. "I cannot perceive anything of that!"

"No, you can't. Your line world is curved in a direction which you cannot observe, since this direction is perpendicular to your world. If the curvature were very pronounced, and your world therefore very small, you would have the fascinating experience of perceiving yourself at a great distance, for your sound follows your world, curving along with a curved route."

"It all sounds like pure nonsense to me," said the king. "Sound following the world. Naturally! How could sound leave it? Curving along, you call it. What nonsense! Inconceivable! And whatever it is that you call curved I can't fathom. Every youngster in Lineland can tell you how it really is!"

"Sound does not travel the shortest route," I replied. "It takes the shortest route possible in *your* world, but actually, the shortest route lies outside your world."

"I no longer understand anything you're saying," said the king.

"That is clear," I told him, "and I can't ask you to understand something which you cannot visualize. I don't think that you have learned much from me, but I did learn a very great deal from you."

"That's certainly not obvious from your statements," the king retorted.

"Just something else you cannot see," I said and disappeared, or rather, Circleland faded away and dissolved into the night.

I awoke, and could just barely recall the vision of Circleland. It had been curved in a direction which was invisible to the inhabitants of that one-dimensional country, because the direction was perpendicular to their space. Now, was this also true of our own two-dimensional world? Could it be that the sides of the triangles which look straight to us are curved in a third direction which is invisible to us? After all, it is conceivable, though we Flatlanders cannot visualize it. How to find that out? The only thing I could think of was to ask the Sphere about it when I saw him again.

And that is what I decided to do.

22 Revelations by the Sphere

I was looking forward in high spirits to my next meeting with Mr. Puncto. From the moment we greeted each other he knew there had to be a reason for my good mood. "Did you find the solution?" he asked.

"No," I replied, "I can't shout Eureka yet, but do believe that I am very close to discovering the basis of the problem. I think that our curious phenomenon—the sum of the angles of a triangle exceeding 180°—has to be explained by assuming that the sides of the triangle are curved but that this curvature is not visible, I mean: *not visible to us*. It occurs in a direction perpendicular to our world. A three-dimensional creature must be able to see the curvature; we cannot."

And I also told him about my dream vision of circle-shaped Lineland which the king could not understand because he could not see the curve.

"It would be very helpful if we could get an opinion about all this from someone living in a world of three dimensions," my friend thought. And that gave me the idea of inviting Mr. Puncto to the New Year's Eve gathering at my home, where we were already counting, perhaps too definitely, on a visit from the Sphere.

When I told this to my wife, she not only agreed but welcomed the idea enthusiastically. While New Year's Eve in our country is a family affair, when all the family's joys and sorrows are recalled, why shouldn't we invite a newly acquired friend too, since we did feel ourselves to be, if not intellectual pariahs, in any case exiles from the world of science?

New Year's Eve was just around the corner. The last weeks seemed like months to me, the last days, weeks—but finally the evening came. My large family circle was present, my wife and I, the children and grandchildren, and Mr. Puncto in the midst of us. His calm, quiet nature, together with his adaptability, made us consider him one of us.

The evening passed delightfully. Of course I told a fairy tale and we ate the customary fried dough circles until it approached midnight and we could expect the visit of the Sphere from Spaceland. It was quite an event for our guest when the Sphere appeared as a tiny circle which grew increasingly bigger until the largest cross-section of our visitor lay in the plane of our world.

After the customary greetings back and forth, I indicated how eagerly we had been looking forward to his visit, all the more so because we were faced with difficult problems which we, Mr. Puncto and I, did think we could solve, but which we should very much like to have confirmed by someone from the world of three dimensions.

The Sphere said he was all ears and I began my story. I spoke of the problem encountered by Mr. Puncto during his trigonometric observations. To stress the importance of his opinion, I elaborated on the consequences all this had had for Mr. Puncto—his dismissal and general disrepute with the scientists and the public at large. After that, I modestly presented the solution we had found for the surprising problem of hav-

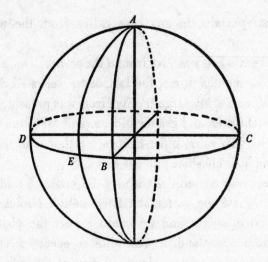

The sum of the angles of a triangle on a spherical surface is always greater than 180°. Triangle ADE has two angles of 90° at D and E; together with angle A they make more than 180°. The angles of triangle ABC are all 90°, their sum is therefore 270°.

ing the sum of three angles of a triangle in our space exceed 180°.

When I had explained our idea about the sides which must be curved in a direction invisible to us, Mr. Puncto thought it necessary to tell our guest that I should be given full credit for this solution—after which it was my turn to give Mr. Puncto a compliment by pointing out that he had understood immediately that the deviation could not be attributed to accidental errors, but must have a special geometric origin.

Then we waited to see what the Sphere's opinion would be. There was quite a long silence. I understood that our guest was busy working out a line of thought with which to clarify the problems, which were of course harder to grasp for us two-dimensional creatures than for an inhabitant of a three-dimensional space. It was flattering to have him start by saying that the description of our problems and the explanation of our solution were perfectly clear to him.

"And what's more," he said, "your solution is correct. The last time I was here in your space, I had already thought of telling you a little more about the form of your space, but thought I might thereby be violating your geometric concept needlessly. However, now I will tell you: your world, your two-dimensional world, is not flat but curved. Let me just point out that in addition to a flat plane, curved planes are also conceivable, even if you cannot visualize this with your two-dimensional observation powers."

"This we do understand," I interrupted him, "because a resident of Lineland cannot know either whether his world is a straight line or is curved, since he is unable to observe the direction in which his world is curved. And so it is actually possible that his world is really a circle and therefore not infinite but finite—though without limits, that is, endpoints."

"Entirely right," the Sphere resumed, "and in the same way your world, as I discovered some time ago, is not straight but curved. You are not living on an infinitely large, flat plane but on a spherical surface."

"Could you explain that a little more fully, please?" Mr. Puncto asked.

"Certainly," the Sphere continued calmly. "You know and understand that a circle is a surface of two dimensions bor·

dered by a curved line, a circular line of one dimension. Similarly, in Spaceland a sphere is an object of three dimensions, bounded by a surface, a spherical surface of two dimensions."

"Therefore," Mr. Puncto said, "if I understand it right, our space is curved, curved all over, without our being aware of it?"

"Right," the Sphere answered, "just as a circle is curved all over without its being noticed by a Linelander living on his circular line."

"But we did notice the curvature of our space, after all," I ventured to point out.

"Yes and no," the Sphere replied. "You will never be able to notice the curve itself, but you did become aware of one of its consequences. Because your flat space is curved and not flat, the sum of the angles of a triangle does not equal 180°. This you observed and it led you to a conclusion which I find brilliant. But precisely because the curve cannot be observed, you will probably continue to remain alone with your conclusion. The common man, or even someone with a scientific background, will not be able to understand you and so will reject it. Only mockery and ridicule will greet you, I'm afraid, when you present a conclusion which is so bizarre to your countrymen. They will question your mental powers. They may obligingly hear you out, but the moment you're out of earshot they will speak sympathetically of senile deterioration or advance the opinion that your mental powers have been seriously damaged by overexertion."

"But," I ventured to bring out, "I am simply going to have to tell the truth. Not only because everyone who has discovered something as yet unknown to anyone else has an urge to make this public, but in this case we also want to clear Mr. Puncto of all the unfair charges which have been leveled against him.

"At first he was suspected of having given the wrong information, and the only result of all our efforts to eradicate this opinion was that at least a willingness was shown no longer to believe that Mr. Puncto had been acting in bad faith; in place of this, however, came the belief that he was an extremely poor observer, that his results were incorrect, and that he was trying to justify himself with the most fantastic prevarications. As far as I am concerned, some people may think my friendship with the victim of these charges was why I risked my good name to take his side—mere courtesy, so to speak— and that there could be no other reason. What else could they think? After all, all I did was to insist that the results Mr. Puncto obtained from his measurements are not based on observational errors. I did talk about triangles which are curved in a direction that is invisible to us, but that was an unexplained hypothesis. Now we can explain it more clearly. After giving the problem much intensive thought, we can clarify this curvature, so inconceivable to us, by means of the Circleland analogy, where we can clearly see the curve of one-dimensional space while the inhabitant of the curved world cannot understand that his line is curved in a direction he cannot see. Finally, I can call on the witness of my visitor of Spaceland who has seen the curvature of our plane."

Here I hesitated, for I realized suddenly that I had been carried away by my enthusiasm. I would never, not even in the faculty, be able to speak about these strange meetings. They would think I had lost my mind or that I had connections with creatures from the spirit world—which would amount to the same thing as having some association with the "Evil One."

The Sphere had obligingly heard me out. At the end he gave me his opinion in a few brief words. He advised us most emphatically not to try to make everything we knew public. As a matter of fact, he even thought it fatal to spring these startling theories on the scholars and scientists of our world. They would certainly not accept them and would write them off as the products of our diseased brains.

And after a brief farewell he left us alone with our problems.

23 Problems

When the Sphere had left us we remained silent for a while. Each of us was occupied with his own thoughts, which were probably running parallel, for we all looked equally somber. And there was good reason for this, since, although we knew our hypotheses to be well founded, we could not expect to make others share our ideas with equal enthusiasm. We stood strong, of that we were convinced, but strong in isolation—and we didn't know whether that would be enough.

The faculty would have to be the first to be won over to our point of view, but these scholarly individuals were the very ones who seemed least receptive to our ideas, and would the arguments, so clear to us, make any impression on others?

Could we expect our judgment to be placed above that of the faculty? Of course not. Not that the minority is necessarily considered wrong, but against the authority of the learned body—which could naturally be expected to issue a public pronouncement—our opinion would not carry much weight.

I proposed that we think the matter over quietly and not take any rash steps.

"But," Mr. Puncto said, "the men of science will have to discover in the end that—in order to explain the strange deviations discovered during the measurements—a hypothesis will have to be drawn up. After all, it is the only scientific standpoint possible. These phenomena cannot be left unexplained."

"I agree wholeheartedly," I replied, "but the matter is not as simple as it seems. Certainly, one must sometimes resort to strange hypotheses in order to explain strange phenomena, but if we put ourselves in the place of these scholars, is the acceptance of those extraordinary conclusions so urgent and self-evident? For the time being, we alone are familiar with the results of your measurements. The deviations are so small, they carry conviction only for someone who has spent sufficient time studying them, and they are so unusual that, whenever possible, one will want to attribute them to observational error."

"That's what we did at first," Mr. Puncto said, "but since the errors occurred with every measurement and since we set our instruments up differently each time, we finally had to believe in them."

"Right," I said, "but try and get someone to devote that much time and energy to something for which he does not see the need or the purpose. Science searches for an explanation of observed facts, but looks only for the simplest hypothesis to explain the facts. And it is simpler to assume that two individuals are—to put it gently—'biased' than to accept a complicated concept of space which is almost incomprehensible to the average person. No, really, their rejecting attitude is the only scientific stand, in their own eyes at least, that they can assume."

Mr. Puncto, my colleague in science and stress, had to agree and we realized that we would simply have to leave it at that. My wife, cautious as wives always are, had not been able to understand our drive from the very beginning. A wife, no matter how closely she shares her husband's inner problems, considers his urge to make his knowledge public as so much intrusion on the outside world.

"I'd keep all this to myself if I were you," she thought out loud. "Why bother others with it if they feel it is of no use to them and if it only leads to more problems?"

Even though we did not really agree with this all the way, we had to admit that it was the best we could do under the circumstances. At least for the time being!

"If we only had one man," Puncto resumed, "one man with scientific standing, who saw the truth—we would be that much further ahead."

"That one man we do have," I cried out, "the Sphere, but for the present let's not ask him to make public demonstrations because we would simply be put away that same instant on charges of black magic and association with the Devil. The knowledge that the Sphere shares our insight helps us and us alone—not anyone else."

My son, who had been listening attentively, interrupted to say: "I would still have liked to ask the Sphere whether the three-dimensional world is also curved in a direction perpendicular to three-dimensional space—in other words, in a direction which a three-dimensional creature is unable to see."

"It's good that you did not," I retorted. "The Sphere is pretty bright because he sees things in our space which we cannot see, but whether he is bright enough to find out if his own space is also curved, something which he cannot possibly see himself, I am not so sure. And I am afraid that we would have annoyed him with such a question. Fortunately you kept the question to yourself."

"I had no time for it because he left so suddenly," my son replied, "but I would like to have that question answered correctly once and for all."

"Later, perhaps. We mustn't antagonize the Sphere. We need his help badly. Very badly."

It was very late, almost morning in fact, when we decided to put an end to it. Mr. Puncto took his leave and we went to bed—but it was a long time before I could fall asleep.

24 The Shortest Way

It was late when I appeared in the family circle the next morning and my mood left much to be desired. My little

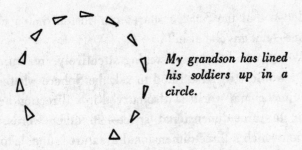

My grandson has lined his soldiers up in a circle.

grandson, on the other hand, was in fine spirits. He sat in the hall playing with his little soldiers, which are small figures in the form of sharp isosceles triangles. An equilateral triangle is the commander.

Eleven soldiers were marching in a circle with the commander at their head. At the order "Forward, march," from the commander, the entire troop started to march. My grandson began by pushing the commander, then the first man, the second, and so on.

Whether I simply wanted to show my interest or whether I wanted to turn my thoughts—still continuously occupied with space problems—in a different direction, I do not know, but I asked him why all the soldiers were not marching ahead simultaneously. I thought the boy would answer that it was impossible to move all his little men at the same time. And I was therefore very surprised to get an entirely different answer.

"They don't hear it all at the same time," he explained.

Now I was really interested. The boy imagined that the sound of the commander's voice moved past the row of soldiers. A rather uniquely scientific observation for such a youngster, but after all, he did come from a family of mathematicians and physicists!

I began to enjoy this and wanted to see if I could trap him. And so I asked: "Does the sound travel to the soldiers along the line of the circle in which they are moving and not along straight lines?"

That brought a moment's hesitation. Obviously, a decision had to be made. But then it came, short and resolute: "No, sound moves along the path they travel. Inside the circle there is nothing."

Of course he simply assumed the latter because he wanted it that way. He wanted to keep things simple. His soldiers were marching in a circle. Everything took place in that circle and to have the sound move inside of it along various straight lines would simply complicate things needlessly.

But my own thoughts were now moving in another direction. I saw a picture of Circleland where the king was standing between his subjects. His circle is his space and his voice moves forward in his space, that is to say, it does not follow the shortest route because that lies outside his space! And so my recent dream popped suddenly back into my thoughts. The king of Circleland was not able to understand me because he did not know there was anything outside his space. It wasn't stupidity which made him unable to comprehend what I was saying, but he simply could not imagine there being anything outside his space. For him, therefore, sound moved along the shortest way possible in his space. We did learn at school that the shortest way between two points is a straight line, but in reality we do not always mean the straight line when speaking of the shortest way.

Suddenly I remembered something from my childhood. I had come home late and my father thought this called for a lecture. He asked me whether I had taken the shortest way home. I had not, but I preferred not to admit it outright. To lie was against my nature and so I looked for a subterfuge, a loophole, and I said: "Of course I did not come home by the shortest route, because that is impossible."

"Why impossible?" my father wanted to know.

"Well," I said, "the shortest way is the straight line and that goes right through houses and all sorts of obstacles."

I have mentioned before that we are all members of a mathematical family and so my father could appreciate my comment.

And now I was again occupied with the problem of determining what really *is* the shortest way. Suppose that the commander of my grandson's little soldiers ordered one of his men to approach him via the shortest route—would the man cross the middle space in a straight line?

It could be that the commander was so busy thinking about the circular formation that he fully expected the soldier to reach him via that route and no other. Furthermore, there might be an obstacle in the middle, a building for example, or simply forbidden territory. Then the commander would certainly not want his man to use what was mathematically the shortest way, but rather the route which was shortest in the practical sense—the circular arc.

But now back to Circleland, where the same holds true. The king knows nothing of a "territory" in the middle. No possibility exists for him other than moving within his own space. That a shorter way is conceivable outside of his space, he cannot know. Perhaps a very brilliant mathematician in Circleland might suspect it, but he would have a difficult time convincing his fellow space citizens of it! Everything belonging to that space cannot leave it. Not even sound. It must follow the curve of the space. Strange really, very, very strange!

I was weary, and it was tiring to go over these questions. I dozed off and dreamed. But, surprisingly enough, this time I did not see a picture of Lineland—the place which I used to visit as a much wiser Flatlander, bent on telling its citizens the truth, so obvious to myself since I could see the real relations which the people were unable to perceive—no, I dreamed something quite different. I was a Sphere from the country of three dimensions and I was visiting my own world, my own Flatland. No, not Flatland, but Sphereland, because I could see clearly now that my world was curved in a direction which had never been visible to me before. But now I could see it, now that I was a Sphere, a wise Sphere! Suddenly I could understand it all so well—but of course this was no great feat, since I was now a three-dimensional creature.

There was my home! I saw it from above, from the out-side—in other words, as I could never see it before. There was my wife. My children! My grandchildren! There, no— how utterly amazing. There I was myself, a pitiful two-dimensional hexagon. How important I had always considered myself to be, and how very insignificant I, a flat, two-dimensional creature, really was!

Should I go to myself? Should I tell myself how everything *really* was and how it looked from the outside of Sphereland's space? Oh no! After all, I'd never believe it anyway —I was too stupid for that. No, not too stupid, but too limited by my faculties, which could only get two-dimensional images of everything around me.

No, don't disturb that learned individual. He just sits there daydreaming a little about problems that are actually much too complicated for him! Leave him be! I shall now look at the two-dimensional world myself, my own world, my own, curved Flatland. Later, when I awaken again there below, will I remember everything I saw? Oh, I hope so! It is so beautiful! It is all so very extraordinary! I see the entire city with its houses, its streets, the trees. Traffic on the street. I see everything alongside everything else. Fascinating, how is that possible in a dream? For I know that I am dreaming. Beautiful! Strange!

I look to the left, to the right, to all sides, and still the world, my world, does not stretch infinitely in every direction. Of course not, because my world is not infinite. It does not stretch out on all sides into infinity. It is a bend, a curved world. I can go around it! I can go and fly all around my world, my spherical Flatland. How utterly fantastic. How can I understand it all so easily? In my dream, did the Sphere provide me with some special insight? Might I then really be seeing with the eyes of my friend the Sphere? Possibly. Who can say?

Yes, how privileged is the Sphere that he can see everything like this. But on the other hand, it does not take much skill. He *sees* it, I have to *visualize* it, I, a Flatlander, even though I am really a spherical Flatlander, a Spherelander. The Sphere does not need to imagine it, he can see it from the three-dimensional viewpoint.

But how about those light rays? How do they travel? Straight ahead? In straight lines? No, of course not. They cannot leave the space, their space. They have to follow the curve of that space because they belong to it. To us, Spherelanders, they look like straight lines. We think that light travels in straight lines. And they are not really curved in our space, but they follow the curve of our space. They have to. But seen from the outside, they are not straight lines, they are nothing but the shortest connecting lines possible on the surface of a sphere. Exactly like the lines along which sound travels in Circleland. Those are not straight lines either. But then . . . if the Sphere is as clever as I am, he must realize that if his space is curved, if his three-dimensional space is curved in a direction invisible to him, then the light rays in his space are not straight either. They will be the shortest lines possible in his space. The rays will have to follow the curvature of three-dimensional space. "The Sphere will have to realize that. If he does not, I am much wiser than he is."

This last I said out loud. I heard myself say it, but it was ill fated. Suddenly I heard a voice, the Sphere's voice, which said reproachfully: "How dare you, flat creature, to place yourself above me. I am three-dimensional and I look down on your world. Ungrateful braggart that you are. Back in your own little world! I let you take a look at your curved world from the outside, but it has gone to your head. Now you belittle me and make me out to be a creature of no intelligence who cannot see the curvature of his own world. Right, I cannot see this curvature, for the simple reason that it does not exist. My world is straight. Straight, do you understand? You are living in a curved world, not I. And now back into it, back!"

I awoke from my dream. Did I dream it, or had the Sphere really brought me outside my own world? Had he let me have a look at my curved world? How could I find an answer to that question? If it were so, how terrible it would be! I would have insulted the Sphere, my good friend and teacher, and he would never visit us again on New Year's Eve. Just as roughly as he had broken off relations with my grandfather, he would now want to have nothing further to do with me. I couldn't let myself think of it!

25 Distant Views

Life resumed its normal course. We were wise enough not to fight for our convictions. The world was not interested. And why wish for fame when it could apparently not be achieved anyway? Perhaps later, after many years! Perhaps one day a new era would dawn in which the world would be more sensible.

Still, it was not easy for either of us, me or my friend Mr. Puncto. When one is convinced of the truth of something, one wants others to benefit from it as well. But no one could appreciate it. There was nothing more to be done. After all, it is not very nice to feel that everyone regards you as a misguided maniac.

We therefore lived quietly, but interest in space problems was in our blood and so it happened that we were among the first to visit the space station which had been constructed high up in the atmosphere. This was something new, since entirely new concepts had had to be applied during its construction. As is known, our houses stay where they are in spite of a small amount of gravity which is pulling everything toward the center of our world disk. It is not very easy to explain how these houses are fixed in place, and I have talked about it once before. But until now there had been no successful attempt to build a stable house at a very high level where the air is thin—this was made possible by a recent technical invention.

A second difficulty was involved. All material had to be transported to a point at high altitude. For this purpose a catapult mechanism had been erected from which everything was shot off in the right direction and at the right speed. The personnel on location were trained to catch all objects arriving on the scene at a very slow speed. Now the building was completed, and even though it was not yet open to the public, those who were especially interested could request permission to visit it. And therefore, when Mr. Puncto proposed to have a look, I was ready to go along. My wife wanted to come too, and the three of us presented ourselves at the appointed time at the catapult-elevator.

The passenger projectile was a comfortable cabin with two compartments, one for the men and one for the ladies. There were four places in the first one. In addition to Mr. Puncto and me, a member of the staff also took a seat. He would be our guide. Of course he had nothing to do while we were going up, but it can happen that something goes wrong and then it is good to have a calm, experienced official in the machine. Let us not even think of all the things that could happen—it would only worry the passengers needlessly. At the very worst, there could be a collision with some object, but fortunately the possibility is very small. There is nothing with which the machine could collide. At most one could imagine that a projectile launched by another station had entered the wrong course. It could also happen that as a result of an explosion somewhere below, fragments from a factory might

The passenger projectile.

get thrown up into the air, but it would have to be a tremendous explosion, which would furthermore have to occur at exactly the right moment. More possibly, the power chosen at the start might be too much or too little. When I asked the guide about this he answered laconically that the engineers were so equal to their task that an error was impossible. I stopped my questioning, thinking that the chance of being part of an accident is always very small anyway.

The compartment for the ladies was next to that for gentlemen. It is clear that it would be too dangerous to transport women in the same compartment as men in such a machine, which receives a shock at the moment of launching. It would certainly have bloody consequences. The little compartment set up for the ladies is very narrow so that the passengers cannot turn themselves. The walls on either end have springs which absorb the shock. An entire bundle of women can be put into this cabin.

The launching shock was not too bad. Our gravity pull is not great, which makes it unnecessary to give the projectile a tremendous starting speed. When the signal for departure went up we did hold our breaths because one is always a little anxious the first time one experiences something like this, but the spring-filled floors of the cabins absorbed the shock quietly. The trip itself seemed to last quite a while. I did not see the end station until we were almost there and a

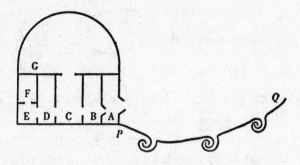

The observation station. (PQ safety net, A airlock,
B *canteen,* C *conversation room,* D *reading room,*
E *and* F *service rooms,* G *observation chamber.)*

few moments later we were lying quietly in the elastic safety net and could get out. It was cold at this height and the air was thin, and we were therefore happy to pass through the airlock into a heated chamber which had a more normal air pressure. The building is completely closed. It consists of two stories. There are several rooms downstairs, a reading room, a conversation room, and a canteen; upstairs, there is the lookout hall, or observatory. It is a very large room, closed off from above by a dome of transparent material. Here many people can enjoy the fantastically beautiful view at the same time.

We stood for a long time in ecstasy, enjoying the magnificent view. All the unpleasantness from the world below was forgotten here. One looked into the unfathomable depths of the universe! My thoughts stopped short. The fathomless depth of our universe, *our* limited universe, which is certainly curved in a third direction invisible to us, but which—meanwhile—remains a limited universe.

We turned our looks to the many worlds floating there in space, worlds like ours, other "universes," as one likes to call them in a dignified way. Some were quite close, clearly visible as small lines, others were far away enough to be seen as small dots, and the small dots which were very far off appeared hazy, because there is still a certain amount of light absorption in empty space.

A lady nearby asked the guide, who was showing us all the beauties of the universe, whether the other worlds were inhabited and what kind of beings might be living there. Businesslike and to the point, the official answered that nothing was known about it, but that this observatory had been built specifically to answer such questions. Observations would be made here, and he predicted that in the coming years the newspapers would be full of the results of this investigation.

"A handsome station, later to be used as exit point for space trips," my wife declared.

"Yes," I said. "Perhaps we shall live long enough to see it. For the time being our catapults are not powerful enough to reach the closest 'universe.' "

Mr. Puncto was not listening but seemed to be deeply preoccupied with his own thoughts. It turned out these were in another direction, because he suddenly said: "If space is curved, we should be able to see ourselves, that is to say, our own universe, in the distance."

"Where?" my wife wanted to know.

"I don't know," Mr. Puncto said. "One of those worlds, very far away, which we here see as hazy dots, is perhaps the same as the one on which we are living."

"Would we then really be able to see our world?" my wife asked enthusiastically.

"Of course it would be possible," I said, "if the distance is not too great, in other words, if our space is not too big."

"Funny terms, but entirely correct," Puncto said. "If our space is not too big. The fact is that our space is not endlessly big."

"Hush," my wife said, "don't let anyone hear you, because they will think that we are not completely normal." And she looked worriedly about her.

Yes, it had come to that. We had become shy in public, afraid of being recognized and pointed out as fools who held crazy notions.

I wanted to divert her attention and asked loudly: "How far away from us would you think those worlds to be?"

Our guide, who thought the question had been directed to him, answered: "That is not known. Measurements will be made at this station, but whether they will produce any result is questionable, because the distances are very great."

26 Telemetry

This last comment had apparently made a great impression on Mr. Puncto—from the moment we were headed back he kept talking about the possibility of telemetry. Our guide, who was also returning with us, knew nothing about it. He had not the slightest idea how such measurements could be made.

"It does not seem feasible to me," Mr. Puncto said, "to measure the distances to far-off worlds from this one fixed station. In order to establish a distance, one has to observe the object involved from two points. In other words, one has to have a base line. From the length of that line and the two adjacent angles, it is possible to calculate the distance."

"That's obvious," I replied, "it is the only way to determine the distance to an object that can't be reached."

"Of course," Puncto resumed, "I only wanted to say that this station is not adequate for that purpose."

"But the observation chamber is quite large, after all," my wife said.

"Certainly," Puncto observed, "but the distance that can be plotted here is extremely small compared to the enormous distances to all the observable worlds."

"But can't it be done in some way from only one point?" my wife suggested.

"Impossible," I immediately told her.

"But," my wife persisted, "when we were up there, you pointed out that the more distant worlds seemed hazier than the ones closer by because there is a light-absorbing medium in space."

That was an astute observation and I admired my spouse for making it. I had to admit she was right, but I explained that this method of measuring was much too inaccurate and

178

that the only reliable way to obtain dependable results is by means of trigonometry.

"Then they'll just have to build a second station," my better half concluded.

"True," Puncto said, "but they are already happy with this one observation station and the authorities would be quite taken aback by a request to construct a second, similar one nearby."

"The second station would not have to be that big or expensive," my wife retorted, and—well—she really had a point there too.

Mr. Puncto had grown very quiet, but I discovered the reason for it only later. He was busy working out plans for long-distance measurements on a large scale. Many weeks went by without our seeing him, and as a result we remained unaware of the preparations with which he was quietly going ahead. Not only did he make the rounds of various authorities, but he also sought contact with the faculty. At first his reception there was anything but friendly, but his proposal did arouse interest and even sympathy on the part of the learned gentlemen. More extensive and accurate telemetric measurements were in the offing for science and the faculty were happy to lend their support to the plan. The government, which had at first objected to what it called the needless expense of a second space station, was finally convinced that it would serve a scientific purpose—and, so far as the expense was concerned, it was not as bad as they thought, since the second space station required only a very simple construction.

Furthermore, as a result of the experiences acquired in building the big station, the construction costs were considerably below the original estimates, and in addition, the regular visits from the paying interested public brought in so much cash it began to look as if the costs of the large, expensive observatory would be covered in a short time, and part of the money could then be set aside for the construction of the small observatory.

Only when everything was "in the bag," so to speak, did Mr. Puncto stop by to bring us up to date on his efforts. We were especially surprised to learn that he had been put in charge of the measuring processes. This had been done not so much to make amends, but because Mr. Puncto happened to be an exceptionally accurate surveyor. There was no danger here that he would be subject to another strange whim and try to explain with bizarre theories the small deviations he would be finding, since only distance determination was involved. Only two angles were measured in each triangle, so there was no danger that he would again discover that the third angle, together with the two others, did not yield exactly 180°.

It goes without saying that he was in the best of spirits and he started his preparations for the construction of his small observatory with high hopes and courage. When it was completed, my wife and I were the first to be allowed to visit it.

Like the large one, this building consisted of two stories. On the first floor the offices for the director and his personnel were located and above that was the observation area. This was rather disappointing at first glance, not only because of its size, but also because there was no transparent dome. Puncto explained that it is better for the observations to be made in the open air rather than through the walls of a light-transmitting roof. The computations were carried out in his workroom below, which had central heating and normal air pressure.

We could understand this very well but it was disappointing to my wife, who had counted on a beautiful viewing dome.

We also visited the other station. There was no connection between the two observation posts—something of an inconvenience for the observers, but it simply had not been

Distance measuring through trigonometry. (The base line AB is known. The angles at A and B are measured. From this the distance of a world C or D can be calculated.)

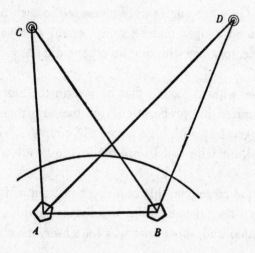

feasible to set up a separate catapulting establishment. And so one had to come down first in order to let himself be catapulted upward again.

The other station had been built next to the great viewing dome where one could make quiet observations without being disturbed by tourists and vacationers.

And so Mr. Puncto started his new work. Together with his assistants he measured the distances of the various worlds, charting the space, as it were. He revived and became a different person, a man who knew himself to be gainfully employed in a profession he loved. It looked as if no one remembered his former mistakes, and he was held in high esteem once more.

27 Increasing Distances

At first our friend Mr. Puncto would stop by regularly, but after a few months he suddenly stopped his visits. He was not ill, for I would run into him occasionally, but he was always in a tremendous hurry and had no time for conversation. Since it does happen that a person is terribly busy for some time or that he is pushing himself to finish a certain amount of work, I did not press him to drop by again. The time would come of its own accord, I thought, when he would feel the need for an intimate exchange of thoughts in a home-like atmosphere, and since he knew he was always welcome at our house I expected to see him there again in time. But when my wife told me one day that she had run into Mr. Puncto and had been struck by how bad he looked and how nervously he was acting, I decided to look him up and inquire about his health and work.

It was not difficult to force a meeting because I knew at exactly what time he passed a certain spot every morning. I pretended to be a little surprised when I met him. Since I stopped right in front of him, he could not move around me without saying something, and so he was forced to reply to my rather obvious questions of "How are you?" and "How is it going with your work?" With a brief "Fine, but very busy," he wanted to be on his way again. I was also struck by the fact that he seemed to be very shy and wary, and before I knew it I had asked him: "There haven't been any more problems, have there?"

His answer, "Yes, no, not at all," made me think. There had to be something and I stood looking after him, deep in thought. My wife had really been right. There was something, something was wrong. We talked it over at home and made some conjectures. Was Mr. Puncto having difficulties with the authorities or with his personnel? Had the old rumors been dug up again and were they making the rounds once more? But all these conjectures did not bring us one step further.

"We should urge him to drop by," my wife said.

"He knows," I replied, "that he is always welcome here, that he can discuss his problems openly with us, and I think he will probably stop by when he is unable to solve his difficulties himself."

Nevertheless, my wife felt that we should tell him this again ourselves. If he did not want to accept our invitation to come and talk things over, we would let it rest, but perhaps he just needed a small push to help him make up his mind to share his problems with us.

I have never found it easy to offer unasked-for help and I was reluctant to present him crudely with an open invitation to reveal his innermost secrets to us. Fortunately, my wife was inclined to be more drastic. One day, when she ran into Mr. Puncto by chance, she said: "Mr. Puncto, we haven't seen you in ages and we would like so much to talk again with you."

"Is your husband having some difficulties?" Mr. Puncto asked.

"We are worried that you may be having them," she replied.

"Do I look as if I do?" he let slip with obvious concern.

"We have been talking about it with each other for some time now," she said, "but you know how much interest we have always had in your work and also in you yourself. Can you drop by for a little while this evening? There will be no other company." And without waiting for an answer she walked on.

She had really managed it very skillfully. That evening Mr. Puncto came by and I greeted our old friend warmly: "We are so glad that you have come. You know how much we have appreciated your visits and how eager we always are to talk all sorts of things over with you."

"I do know that," he replied, "but I am afraid that I am not very good company. Just half a year ago everything was cheerful. We had all been revived—but now it is different. I am afraid to dampen everyone's spirits here."

"But have you forgotten," I reminded him, "that we also used to discuss all difficult problems in the old days and that we did a rather nice job of solving them?"

"Yes, but haven't you also forgotten," he countered, "how I dragged you with me through all that misery? We were regarded as the town's prime pair of fools, a couple of dilapidated maniacs hawking the wildest ideas. I cannot and must not drag you through it again."

"Is something really serious the matter?" I asked, and the expression on my face must have shown how greatly interested I had suddenly become.

"It's serious. Something is wrong again," was his somber reply.

"Then we will have to devote our best powers to it," I said, determinedly.

"In any case, my husband and I will certainly not leave you alone with your problems," my wife added.

"I know that you will help me," Mr. Puncto said, "and you must forgive me. I really don't know where else I can turn."

"Dearest friend," I comforted him, "we would be very hurt if you did not share this with us."

After that our guest gave in. He began to tell us how in the beginning he had trained his personnel for the observations which had to be carried out at the two stations. In connection with this, two difficulties cropped up which were overcome after some concentrated effort—namely, the observations had to be made simultaneously, and furthermore, the observers at the two stations had to have the same object in view. If the observers were to sight their instruments on two different objects the result would be wrong—it would show the distance to a nonexistent object.

After a series of practice sessions everything went smoothly. In order to control the observations, the director sometimes had the distance to the same object measured again, and the results always tallied closely.

Finally the time had come to start work in earnest. The location of many worlds, some fairly near, others farther away, and still others very far off, was determined and much progress was already being made toward the charting of the universe as a whole.

After several weeks of intensive work, the director granted himself and his staff a few months of vacation after which everyone resumed his work with renewed ardor.

To get a proper start, Mr. Puncto first had some earlier observations repeated. The distances to a few worlds, located at different ranges, were once again measured for comparison with the earlier data. Nothing tallied!

What had happened? His first reaction was to doubt the instruments, but he was unable to find a reason for the deviations there. Then he repeated all his earlier observations and compared the results carefully. It was true, all the distances were changed, they had all increased, some considerably, others less so. What could be the reason? How to explain it?

Not that all discovered values were changed, but how did it happen that all the established distances had been modified according to the same pattern? It goes without saying that Mr. Puncto, who had already had a bad experience with the authorities, was afraid of being regarded once again as a bad observer, a bungler, or even a cheat. And this disaster could no longer be circumvented. The faculty, which was supervising the research, wanted to see the figures. Then it all came out into the open and the conclusions drawn by the learned gentlemen were not very flattering.

The city government did not want to act right away on the faculty's suggestion that they dismiss Mr. Puncto and instead sent him off on two months' sick leave. His personnel was put to work elsewhere and the observations came to a halt.

Following the rest period Mr. Puncto resumed his work, anxiously wondering what the results would be this time. And it was saddening indeed that he found even greater discrepancies. It almost seemed as if the worlds were rushing away from us in every possible direction. But was that conclusion acceptable? In any case, it looked odd.

These were the problems Mr. Puncto had been telling us about. It was, of course, an interesting case for unbiased investigators, but for Mr. Puncto the affair had another side. His employers, the government, supported by the judgment of the faculty, did not believe him, and his findings had placed Mr. Puncto, who did not have much more to lose, in a very unfavorable light with them.

28 *Hunting for the Cause*

"There has to be a cause," was my first reaction.

"You're a little hasty with your conclusion," my wife said.

"But I am sticking to it," I said. "This has to be our point of departure. There has to be a cause. It would be too much of a coincidence if all distances simply increased without reason. When a marksman aims at a target, not all his shots hit the bull's-eye. Most of them fall wide of the range—to the left or to the right. There is just as much chance of his going too far to the left as to the right. It's true that even with a large number of shots not exactly as many will fall to one side as to the other, but the difference between the numbers will be small. Now, if it turns out that out of some one hundred shots not a single one hit the target and not a single one fell to the right, but all of them went too far to the left, everyone will realize immediately that there must be a reason. For example, the gun may be faulty or the wind may have been blowing from the right."

"True," said Mr. Puncto, backing me up. "There has to be a reason, but what is it?"

"Could a very strong wind be blowing those worlds away from us?" my wife wondered.

"Impossible," Mr. Puncto replied. "The higher we go, the thinner the atmosphere becomes. We have to assume there is no atmosphere between the worlds. There can be no question of their being blown away from each other."

"I understand that," my spouse resumed, "but I can't see any other reason."

"We can't either," I said again, "but there undoubtedly is one. There has to be. I am convinced of it, and Mr. Puncto is too, I believe."

"Yes," Mr. Puncto answered. "I did not dare say it, but I agree with you wholeheartedly. I could not very well come up with that idea myself because it would seem I was looking for a way to get out of the responsibility for making faulty observations."

"There can't be any question here of faulty observations," I declared. "I know you for an accurate observer and I am convinced that there has to be a natural reason for this phenomenon, this dispersion of those worlds."

"But we can't just push it off on natural laws," Mr. Puncto resumed, "without specifying what these laws are—and I see no way of doing that."

"Oh, but we cannot be sure of that beforehand," I reassured him. "We have to look for it and try to discover the reason. After all, we shouldn't begin by giving up all hope before we have made a serious search."

"I have already thought about it long and hard," Mr. Puncto said dispiritedly, "with no results. Nature's secrets sometimes prove to be unfathomable."

"We will make every effort. In any case, it is an interesting problem," was my opinion.

Puncto sighed heavily and said: "Of course I am most grateful to you for your offer to help me, but I am afraid that this time we will not succeed; this is one instance, I believe, when we will not be able to wrest nature's secrets from her."

"But you will of course help me and give me every possible assistance?" I cried out.

"Now you are turning the matter around," Mr. Puncto said surprised. "I help you? But I am the one with the problems, after all, and now you are asking me if I will assist you."

"Then we will work on it together," I rejoined. "Mr. Puncto, just as we did the last time, we will test our thoughts and ideas on each other. In such a case much greater headway is usually made through mutual consultation than on one's own. We start tomorrow. Agreed?"

"Delighted," Puncto replied, "but . . . start? With what do you want to start?"

"I want to start," I said, "by looking over your numerical material, not to check it for mistakes but to see if the figures can teach us something, if they can give us a clue."

Mr. Puncto thought it an excellent plan and we agreed that I would join him the next morning to go over the calculations and their results. Visibly relieved, our guest departed from our house late that night.

The next morning I was sitting in Mr. Puncto's study as he explained how the measurements had been made and in what way the figures had been processed. Before long I was deeply immersed in the entire problem. It took several days before I had a good general picture, but I finally did, and the moment had arrived to discuss the question in every detail.

Mr. Puncto thought there was nothing more to be discussed since both of us were now familiar with every aspect of the question—and he was actually right.

"But why don't we check," I suggested, "to see whether we cannot make some variation in the observations, in order to find out whether we would still get the same results?"

"What did you think of changing?" Puncto asked. "To assume another baseline would mean asking the authorities to build another observation station, and in view of our very questionable results so far, the gentlemen would undoubtedly refuse to go along."

"But can't we conceive of an entirely different method to measure the distances from the objects in space to our little world?" I ventured to ask. The question seemed superfluous to me, without any prospects, but after all, such a question has to be asked to explore every possibility. It was therefore not directed to Mr. Puncto as such, but to both of us together. While neither of us had an answer, the question continued to be uppermost in my mind all the way home, and I kept asking myself: "Isn't there another way of measuring distances?"

I went through every way that I could think of. To walk the distance to be measured and note the time it took was of course one method, but impracticable in this case. Another way would be to shoot off a projectile and see how long it took to hit the target. Just as impossible here, although . . . I did not want to discard the idea altogether. Wasn't there something else we could use besides projectiles? But what? For the time being I had no answer.

That night I could not get to sleep. Every small detail

from our earlier research came to my mind and suddenly I
saw Lineland again in front of me and heard the king explain
how he determined the distances to his subjects by means of
the time it took for him to receive an answer to his call. He
received the answer right away from those of his subjects
who were standing close to him, but from someone farther
away it took a while, and this helped give him an idea of
the distance. Could distance then not be measured by means of
sound in our space as well? No, impossible, because in our
space sound does not travel through a vacuum to other
worlds. Then what about light? Yes, light! That might be
possible. Still, how could one know how long it takes a light
ray from a distant world to reach us? We cannot establish
the exact time that the light beam left the distant world. So
this would not work either, and tired out from all my thinking
I fell asleep.

At my next discussion with Puncto I gave him all the
ideas that had occurred to me, including distance measurements by means of sound and light, and I said that both
were impracticable.

"Perhaps not both," Puncto said. "It's true we cannot
determine at what time a light signal sent by us will reach a
distant world, nor can we know when a light signal coming
to us from a far-off world left there, but perhaps we could
send one out which is sent back to us when it reaches its
destination. We might then be able to figure the distance on
the basis of the time it took the light to travel that distance
back and forth. After all, that was how the king of Lineland

worked his calculations out. He gave a sound signal and the subject for whom it was intended would immediately send a similar signal back."

I had to stop and think a moment, and an objection occurred to me. I said: "All well and good, but there is no one on that distant world to receive our signal and return it to us. We might assume the other worlds to be inhabited, but even so we would not be able to instruct the inhabitants, with whom we have no contact at all yet, to collaborate with us on a scientific test."

Puncto burst out laughing. "No," he said, "that is impossible, but we could send out a light signal which would bounce off a distant world and return to us."

"All right," I told him, "but how would you go about placing a mirror on that distant world to bounce the light back to us?"

"There are," Puncto explained, "beams which can be sent out by means of a specially designed instrument. They are called radar beams, and are bounced back by any object. They are well suited to our purpose."

This was a novel approach and we decided to apply radar beams to distance measurement. It took only a few weeks to install the new instrument in the observation station and we proceeded with our measurements right away. At the same time we measured the distance to an object in space in the old manner, and the two methods gave the same solution every time. With nearby objects, the old trigonometric method proved to be the most accurate; with remote objects, on the other hand, for which the baseline was too short proportionately for accurate trigonometric calculation, the radar measurement provided a better result. Each time, however, the results of the two methods coincided. There was no doubt: all the worlds were indeed moving further away from us, albeit at different speeds.

Yet we had not come a step closer to the reason for the amazing phenomenon of dispersing worlds. The results could no longer be doubted: we therefore had to discover the reason from those results.

I suggested drawing up a list of all the observed worlds with separate columns for the characteristics differentiating them, and for the amount of speed at which they were receding.

Even though the worlds did not have definite colors, differences in shade were to be noted. Some were more reddish, others yellowish or bluish, with other shades in between. But there turned out to be no connection between the color of a world and the speed at which it was receding.

Remote worlds naturally look smaller than nearer ones, but it is not always true that an apparently large object in space is closer than one which looks smaller. Actual size can be determined from the apparent size if the distance of the object is known, and this actual size was also included in our list; but it too appeared to have no connection with the speed of withdrawal.

However, there was another column which did provide the looked-for concordance—namely, the distance from the worlds to us. It was obvious that the more remote worlds were pulling away from us faster than the closer ones, and there seemed to be no exceptions. The speed of withdrawal increased consistently with the distance. Here, then, was the connection—but how to explain it?

An obvious assumption would be that light is slowing down, that the speed of light is decreasing. We might even consider this assumption if we had limited our distance measurements to light rays. But the trigonometric measurements produced the same results, canceling out the hypothesis. Another reason had to be found.

29 Expanding Circleland

For days I walked around with the problem and I was beginning to doubt if I would ever find a solution to it. Still, my spirits stayed high. It has been my experience that the right insight will sometimes appear unexpectedly and suddenly. One simply has to carry the problem around wherever one goes, be more or less occupied with it all the time, and sleep on it.

I couldn't find any rest. In my room I would march up and down, in my sleep I would turn and twist, when I was outside I would walk without paying any attention to the traffic—all with no result. But I could not let go of the problem, and neither would it let go of me. One day, after returning home from a stroll through the city alone with my thoughts, I found my grandson playing with his soldiers in the hall. They were lined up in a circle, a fairly small circle at that, which did not give them much room to move. That seemed to bother the boy, too, and he therefore made the circle a little larger. Now the little soldiers were standing farther away from each other.

"What are you staring at, Grandpa?" the boy asked.

"Well, I am really thinking along with your game," I answered. "Why did you increase the size of the ring?"

"To make more room, of course," he rejoined. "The bigger the circle the more space there is; the distance between the soldiers has now become greater. That's all."

"No, that's not all," I said. "The officer will now find out that it will take longer before his orders are followed because it takes more time for sound to reach the soldiers."

"Yes, especially those standing farther away," my grandson added. "The farther away they are the longer it will take them to hear commands."

There I stood, I, the professional mathematician! I had to learn the answer to my question from my grandson. At least, his words held a possibility, perhaps *the* possibility. The distances had become greater here because, because . . . yes, because the circle in which the soldiers had been set up, the very world of those soldiers in other words, had become bigger. Their world had expanded. Expanded! Expanded? Weren't my thoughts taking a very peculiar turn? Could it really be, could a *space* expand? Yes, here the circle of the little soldiers had really expanded, but could I apply this to a world, even if it were circle-shaped? As if in a dream I marched to my room, and, tired from the walk or all that thinking, I fell asleep.

There was Circleland again. I saw the king with all his

subjects. The king seemed to be angry. Why? It soon became clear to me. He was taking roll call. Every time he called out, one of his subjects had to answer. First those standing beside him, then those next to them, and so on. He did this as befitted a proper king, not once but three times, and what happened? Every time the answers came back following a longer interval. I could see why. I saw that his space increased, his space, his circle space was slowly expanding. The distances between his subjects gradually increased and, accordingly, a sound signal to one of them took longer to return the second time than the first. Each time it took a little longer. I saw it! His space was becoming greater, but the king did not understand this. He thought his subjects were sabotaging his commands, yet he ought to see this could not be the reason because of the regularity with which the returning sounds were delayed.

Should I tell him? No, better not. I had had enough

Expanding Circleland. The distances between the objects increases.

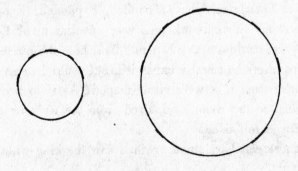

experience with that poor Circlelander. He never did want to accept my explanations. It was not really his fault, of course, he could not understand anything about the expansion of a circle if he had no idea what a circle was in the first place. A creature which can move along one line only can never understand the concept of a curve—for that he would need a two-dimensional concept, at least. All much too complicated for a king of Lineland or Circleland.

But *I* could see it. The distances in his space were increasing because his space was inflating, expanding. That was the reason.

I don't know when I awoke, because my dream slowly turned into wide-awake thinking. Had I been woolgathering or had I done some deep thinking there? I thought of Circleland, expanding Circleland, and now I came back to my own world, my Flatland, my Flatland which was curved in a direction that was not observable to me, my Sphereland. Here too the distances were increasing. Would it . . . also here? Yes, also here . . . ?

30 *Expanding Sphereland*

It was difficult to take the next step. Difficult or only bizarre? After all, it was only one step. As a result of Circleland's expansion, the distances in that world were increasing. Then in the case of expanding Sphereland, the distances between the various worlds here would have to increase, wouldn't they? I wanted to discuss it with Mr. Puncto. What would he say? I ran out of the house to talk the whole thing over with my friend and co-worker.

So we did. Mr. Puncto thought the explanation interesting but very strange. And yes, it was odd of course, very odd in fact, but wasn't that because we were two-dimensional creatures? An expanding Circleland is not that curious to us, but it certainly is for its king, so wise within his own realm, but nevertheless incapable of seeing curvature.

Mr. Puncto walked home with me. We did not talk about it any more and were really not sure where we were walking, both of us being deep in thought. We just strolled alongside of each other, looking at everything and nothing. And so we found ourselves coming to a halt at the gas factory and looking at a gas container which was being filled, without really paying any attention to what we were seeing. Such a gas container is a circular object of ductile material. When gas is blown into it, it expands. I had seen it many times before. Something like this is interesting to a boy and very ordinary to an adult. I was watching it without much interest when something suddenly stirred in my mind. The expanding casing was inscribed with letters which spelled out COOK WITH GAS. At first the letters were close to each other, but the distances between them increased in proportion to the container.

"Do you see that?" I said to Mr. Puncto. "The casing is expanding."

"Yes," he replied, "and the distances between the letters increase."

So he had seen it too.

"Now watch," I said. "Take only one letter. The distances from the other letters to that one all increase. Now suppose that some intelligent creature is perched on top of that one letter observing the other letters. He will see they are all moving away from him—the closer ones at a low speed, but the farther ones at greater speeds."

"That's right," Mr. Puncto exclaimed. "We've got it. That's the solution! In our two-dimensional curved Flatland, our Sphereland, therefore, something like this must be happening too! Our space is increasing, it expands, and as a result the distances between all worlds increase, and we therefore see them move farther away from us."

"Those farther away recede faster than the near ones," I added.

"We have really found the answer!" Mr. Puncto cried out, visibly excited. But he immediately followed with a dejected: "This solution will never be accepted by the professors."

Unfortunately, he was right. When he was called up before the department a few days later to testify about his work, and when he could produce no better results than before, he tried to explain the change in distances by means of our expanding-space concept. He was heard out with cold indifference. It was obvious that the gentlemen thought it all pure nonsense, a way for Mr. Puncto to get himself out of his difficulties. Mr. Puncto sensed the futility of his efforts and was not greatly surprised when he received his dismissal notice at home a few days later. The authorities had not been able to oppose the judgment of the highest institute of learning any longer.

It was a sad affair. We felt that we had discovered the right solution, but what can you do when the world of learned men is incapable of understanding the solution? When we talked the matter over at my home, I said: "Mr. Puncto is a greater man of learning than those gentlemen on the faculty." To which my wife added loyally: "And so is my husband!" But although this was nice to hear, it was cold comfort, especially for Mr. Puncto who had already suffered a similar blow once before. He accepted his second firing with resignation, however. Apparently it was a fate he could not avoid.

31 Miracles in Spaceland

Friendship with my family did our bachelor, who was all alone in the world, much good, and it went without saying that he joined us at home on New Year's Eve, the time when everyone lets all the good and the bad of the past year pass in review while cautiously reflecting on what favorable prospects the coming year may bring.

It was a sizable group which had gathered in the living room and as head of the family I was the center of interest. It went without saying that I would have to tell a fairy tale. I had been doing it for so many years now, and I always found it a pleasant duty. By this time all the well-known fairy tales had been given several turns each—there were no new ones. And yet . . . I had thought of something, not a new fairy tale, but I would dress an old one in a new garment, a new solution for an old problem. And so I began with The Sleeping Beauty.

How did it happen that the forest around the castle, which had remained impenetrable for more than a hundred years, could suddenly become passable for a prince—who was no Isosceles with a very sharp vertical angle?

I found the answer in the expansion of our space, as a result of which the distances between the trees increased. If the trees were farther away from each other, the space between them would have increased. I must admit I was delighted with my discovery and I was quite disappointed when my grandchildren were unable to appreciate this solution. Children become attached to a story in its traditional form and will not tolerate the slightest change in it. I did tell them that distances between points increase on an expanding circle, but when my oldest grandson asked: "But we aren't sitting on a circle, are we?" I realized that he could not grasp the extension to an expanding two-dimensional space. The adults did not understand my explanation either, except for Mr. Puncto, and so it happened that this time only the narrator and his guest were able to enjoy the story wholeheartedly.

Fortunately, the fried dough circles, which were quickly passed around, helped divert attention and a genuine New Year's Eve mood set in.

When the children had gone to bed, the grownups stayed behind to wait for the Sphere's expected visit. And fortunately he did not fail us. On the stroke of twelve he descended into our space. When the customary greetings had been exchanged, the conversation was guided to the burning issues, and after an extensive report on all the events of the past year, I asked our three-dimensional guest what he thought of our solution. To our great satisfaction we turned out to have been right. The Sphere told us our space is an expanding two-dimensional world—not a circle, but a spherical surface which expands at a regular rate.

Being a three-dimensional creature, he could easily see our expansion. He could see, for example, that the distances between all points on the sphere surface grow and he could also see the spots on the sphere surface move away from each other, whereby the distances between the most widely scattered spots naturally increased more rapidly than those between the closer spots.

We were satisfied, but I still had one burning question I dared not ask. My son did dare, however—much to my shocked surprise. He asked whether a similar phenomenon had not been observed in the three-dimensional world. Fortunately, the Sphere did not get angry, but said calmly that this was indeed the case. The three-dimensional universe contains worlds which are called nebulae, since their vast distances cause them to look like tiny, hazy dots. It was observed that these little dots were moving away from each other, and there, too, just as in our Sphereland, the rate at which they withdraw from any other dot increases with distance.

So it is not only possible for a curved, one-dimensional world, Circleland, to exist and to expand steadily—but it holds true for a two-dimensional world, namely our swelling

Sphereland, and even for one of three dimensions, a curved Spaceland which is also expanding steadily. It was clever of the Sphere to understand this, even though he was unable to see it, just as we could not observe our expanding spherical surface.

It was all clear to us now— to us, but we were the only ones in our universe who understood it. We were alone, completely alone!

32 *Misunderstood*

So far as we were concerned, we had reached the apex of our glory, and without self-glorification we did feel that some honor was due us from our fellow countrymen. We soon realized, however, that nothing would come of it. The world is not yet ready for such insights. They could not understand us. As has happened so often in human history, it will probably be said of us only much later: "Those two men were ahead of their time!" Some day a monument may even be raised in our honor, as for my grandfather, who was also honored only after his death.

It seems useless to us to keep sparring with the learned gentlemen at the faculty and to let ourselves be pointed out by the general public as two prime idiots. There'll come a time when it will be realized that we were right, that our concept of things is the only right one. Others will discover the growth of distances too, and in the end an explanation will have to be devised, and since that explanation cannot be found in the assumption that our space is straight or non-curved, the correct concept will have to be formed. Then it will be realized that our space is curved and that this curved space is expanding.

How long it will take for all this to happen, we cannot say. My grandson may still witness the time when his grandfather, regarded as a fool in his own time, is honored as "the discoverer of expanding curved space." Then again, it may well take many generations for that insight to mature.

Life moves on. I have promised my wife not to force my notions on a world which does not want them. I will not try to make the world ripe for them. Acceptance has to come about naturally and so it will. Later. That is why I have recorded everything in this book.

I will let the manuscript be sealed and entrust it for safekeeping to the city librarian. I will paste a label on it with the inscription: "To be opened whenever the theory of the expanding universe has been accepted."

As long as this inscription is considered to be a mere bit of foolishness, the little package will doubtless stay where it is, but should some librarian get to see it at a time when the world's opinions have changed so radically that he realizes the significance of my words, he will no doubt open it. Then, I hope, there will be people, men of learning, with a wider outlook than the professors of the faculty today, who will grasp the importance of the data in the book and then my work will get to see the light posthumously. It is bound to happen.

We never saw the Sphere again. He has probably passed

away, and another official delegate from Spaceland may appear at the new millennium, but I won't live to see it.

Since I am well advanced in years, I am no longer dependent on the goodwill of the authorities. Mr. Puncto, on the other hand, is younger and inactivity goes against his nature. Now that he has kept quiet for a long time, he has been given another position. Of course he will never again receive a prominent appointment in the world of science. He will not be asked to do telemetry in space and he is considered to be unfitted for surveying. Instead he has been appointed as agent for the internal revenue, a job for which he is thought to be eminently qualified. After all, he is known to have an aptitude for figures. Exactness is not so important in his job and in the eyes of the authorities his outlandish ideas have helped make him well qualified.

Obviously he does not have many friends and cannot be received at home by the top-ranking families. At our house he is always welcome, however, and he continues to make much grateful use of the hospitality he finds here.

Index